Evolutionary Biology from Concept to Application

Pierre Pontarotti
Editor

Evolutionary Biology from Concept to Application

 Springer

Editor:
Dr. Pierre Pontarotti
UMR 6632
Université d'Aix-Marseille/CNRS
Laboratoire Evolution Biologique et Modélisation, case 19
3, Place Victor Hugo
13331 Marseille Cedex 03
France
Pierre.Pontarotti@univ-provence.fr

ISBN: 978-3-540-78992-5   e-ISBN: 978-3-540-78993-2
DOI: 10.1007/978-3-540-78993-2

Library of Congress Control Number: 2008924864

© 2008 Springer-Verlag Berlin Heidelberg

This work is subject to copyright. All rights are reserved, whether the whole or part of the material is concerned, specifically the rights of translation, reprinting, reuse of illustrations, recitation, broadcasting, reproduction on microfilm or in any other way, and storage in data banks. Duplication of this publication or parts thereof is permitted only under the provisions of the German Copyright Law of September 9, 1965, in its current version, and permission for use must always be obtained from Springer. Violations are liable to prosecution under the German Copyright Law.

The use of general descriptive names, registered names, trademarks, etc. in this publication does not imply, even in the absence of a specific statement, that such names are exempt from the relevant protective laws and regulations and therefore free for general use.

Product liability: The publisher cannot guarantee the accuracy of any information about dosage and application contained in this book. In every individual case the user must check such information by consulting the relevant literature.

Cover design: WMX Design GmbH, Heidelberg

Printed on acid-free paper

9 8 7 6 5 4 3 2 1

springer.com

Preface

Every biological system is the outcome of its evolutionary history; therefore, the deciphering of the evolutionary history is of tremendous importance to understand biology.

Since 1997, scientists of different disciplines who share a deep interest in evolutionary biology concepts and knowledge have held an "evolutionary biology meeting at Marseilles" in order to discuss their research, exchange ideas, and start collaborations. Lately, scientists interested in the application of the concepts have joined the group.

This book is a selection of what I think are the most representative talks of 11th meeting as they represent an up-to-date overview of concepts of evolution and how these concepts can be used to understand biology in general. The book comprises several topics that we have arranged in different subcategories: modelization of evolution (Yu: Chap. 1, Meade and Pagel: Chap. 2), concepts in evolutionary biology (Toll-Riera et al: Chap. 3, Erenpreisa and Cragg: Chap. 4, Mikhalevich: Chap. 5, Raineri: Chap. 6), knowledge (Shimizu: Chap. 7, Hwang et al: Chap. 8), and applied evolutionary biology (Barthélemy et al: Chap. 9, Hilu and Barthet: Chap. 10, Kryger and Scholtz: Chap. 11, Swynghedauw: Chap. 12, Levasseur and Pontarotti: Chap. 13). I hope that this book will be useful not only to evolutionary biologists but also to biologists in general and that it will help to produce a needed epistemological shift in the different domains of biology—genomics, postgenomics, etc.—from correlative approaches to evolutionary approaches.

I would like to express my sincere gratitude to the other members of the scientific committee—Etienne Pardoux, Philippe Monget, Bernard Swynghedauw, Etienne Danchin, Vincent Laudet, and Michel Milinkovitch, to the organizational committee—Virginie Lopez-Rascol, Olivier Chabrol, and Julie Perrot, to the sponsor—Université de Provence, CNRS, GDR BIM, Région PACA, Conseil Général 13, Marseille Provence Métropole, and to Association pour l'Etude de l'Evolution Biologique that organizes the meetings. I also wish to thank the staff at Springer for their cooperation.

Last but not least I thank the meeting coordinator Axelle Pontarotti for her work and also, as many participants said, for making us feel at home during the meeting.

Marseilles, France *Pierre Pontarotti*
February 2008

Contents

Part I Modelization of Evolution

1. Rate of Adaptation of Large Populations 3
   Feng Yu and Alison Etheridge

2. A Phylogenetic Mixture Model for Heterotachy 29
   Andrew Meade and Mark Pagel

Part II Concepts in Evolutionary Biology

3. Accelerated Evolution of Genes of Recent Origin 45
   Macarena Toll-Riera, Jose Castresana, and M. Mar Albà

4. Life-Cycle Features of Tumour Cells 61
   Jekaterina Erenpreisa and Mark S. Cragg

5. General Evolutionary Regularities of Organic and Social Life 73
   Valeria I. Mikhalevich

6. Old and New Concepts in EvoDevo 95
   Margherita Raineri

Part III Knowledge

7. Overturning the Prejudices about Hydra and Metazoan Evolution . . 117
   Hiroshi Shimizu

8. The Search for the Origin of Cnidarian Nematocysts
   in Dinoflagellates 135
   Jung Shan Hwang, Satoshi Nagai, Shiho Hayakawa, Yasuharu Takaku,
   and Takashi Gojobori

Part IV Applied Evolutionary Biology

**9   A Possible Relationship Between the Phylogenetic Branch Lengths
and the Chaetognath rRNA Paralog Gene Functionalities:
Ubiquitous, Tissue-Specific or Pseudogenes** 155
Roxane-Marie Barthélémy(✉), Michel Grino, Pierre Pontarotti,
Jean-Paul Casanova, and Eric Faure

**10   Mode and Tempo of *matK*: Gene Evolution and Phylogenetic
Implications** 165
Khidir W. Hilu and Michelle M. Barthet

**11   Phylogeography and Conservation of the Rare South African Fruit
Chafer *Ichnestoma stobbiai* (Coleoptera: Scarabaeidae)** 181
Ute Kryger and Clarke H. Scholtz

**12   Nothing in Medicine Makes Sense Except in the Light of Evolution:
A Review** 197
Bernard Swynghedauw

**13   An Overview of Evolutionary Biology Concepts for Functional
Annotation: Advances and Challenges** 209
Anthony Levasseur and Pierre Pontarotti

Index 217

Contributors

M. Mar Albà
Research Unit on Biomedical Informatics, Institut Municipal d'Investigació
Mèdica, Universitat Pompeu Fabra, E-08003 Barcelona, Spain,
and
Catalan Institution for Research and Advanced Studies, 08010 Barcelona, Spain,
malba@imim.es

Roxane-Marie Barthélémy
Laboratoire Evolution Biologique et Modélisation, Case 19, UMR 6632, Université
d'Aix-Marseille/CNRS, 3 Place Victor Hugo, 13331 Marseille CEDEX 03, France,
roxane.barthelemy@univ-provence.fr

Michelle M. Barthet
ARC Centre of Excellence in Plant Energy Biology, School
of Biological Sciences, University of Sydney, Sydney, NSW 2006, Australia,
michelle.barthet@bio.usyd.edu.au

Jean-Paul Casanova
Laboratoire Evolution Biologique et Modélisation, Case 19, UMR 6632, Université
d'Aix-Marseille/CNRS, 3 Place Victor Hugo, 13331 Marseille CEDEX 03, France,
bioplank@up.univ-mrs.fr

Jose Castresana
Institute of Molecular Biology of Barcelona, CSIC, Barcelona, Spain,
jcvagr@ibmb.csic.es

Mark S. Cragg
Tenovus Research Laboratory, Southampton University Hospital; Tremona Road,
Southampton SO16 6YD, UK, msc@soton.ac.uk

Jekaterina Erenpreisa
Latvian Biomedicine Research and Study Centre, Ratsupites str. 1, 1067 Riga,
Latvia, katrina@biomed.lu.lv

Alison Etheridge
Department of Statistics, University of Oxford, 1 South Parks Road, Oxford OX1
3TG, UK, etheridg@stats.ox.ac.uk

Eric Faure
Laboratoire Evolution Biologique et Modélisation, Case 19, UMR 6632, Université
d'Aix-Marseille/CNRS, 3 Place Victor Hugo, 13331 Marseille CEDEX 03, France,
eric.faure@univ-provence.fr

Michel Grino
Inserm UMR 626, UFR de Médecine Secteur Timone, 27 Bd Jean Moulin, 13385
Marseille CEDEX 5, France, michel.grino@medecine.univ-mrs.fr

Takashi Gojobori
Center for Information Biology and DDBJ, National Institute of Genetics, Mishima,
411–8540 Shizuoka, Japan, tgojobor@genes.nig.ac.jp

Shiho Hayakawa
Center for Information Biology and DDBJ, National Institute of Genetics, Mishima,
411–8540 Shizuoka, Japan, shayakaw@lab.nig.ac.jp

Khidir W. Hilu
Department of Biological Sciences, Virginia Tech, Blacksburg, VA 24061, USA,
hilukw@vt.edu

Jung Shan Hwang
Center for Information Biology and DDBJ, National Institute of Genetics, Mishima,
411–8540 Shizuoka, Japan, jhwang@lab.nig.ac.jp

Ute Kryger
Department of Zoology and Entomology, University of Pretoria, Pretoria 0002,
South Africa, ukryger@zoology.up.ac.za

Anthony S.G. Levasseur
Laboratoire Evolution Biologique et Modélisation, Case 19, UMR 6632, Université
d'Aix-Marseille/CNRS, 3 Place Victor Hugo, 13331 Marseille CEDEX 03, France,
anthony.levasseur@univ-provence.fr

Andrew Meade
School of Biological Sciences, Philip Lyle Building, The University of Reading,
Reading RG6 6BX, UK, a.meade@reading.ac.uk

Valeria I. Mikhalevich
Zoological Institute, Russian Academy of Sciences, Universitetskaya nab. 1, St.
Petersburg 199134, Russia, mikha@js1238.spb.edu

Satoshi Nagai
National Research Institute of Fisheries and Environment of Inland Sea, Maruishi
2-17-5, Hatsukaichi, 739-0452 Hiroshima, Japan, snagai@affrc.go.jp

Mark Pagel
School of Biological Sciences, Philip Lyle Building, The University of Reading, Reading RG6 6BX, UK, m.pagel@reading.ac.uk

Pierre Pontarotti
Laboratoire Evolution Biologique et Modélisation, Case 19, UMR 6632, Université d'Aix-Marseille/CNRS, 3 Place Victor Hugo, 13331 Marseille CEDEX 03, France, pierre.pontarotti@univ-provence.fr

Margherita Raineri
Department of Biology, University of Genoa, Viale Benedetto XV 5, 16132 Genoa, Italy, raimrg@unige

Clarke H. Scholtz
Department of Zoology and Entomology, University of Pretoria, Pretoria 0002, South Africa, chscholtz@zoology.up.ac.za

Hiroshi Shimizu
National Institute of Genetics, 1111 Yata, Mishima, 411-8540 Shizuoka, Japan, hshimizu@lab.nig.ac.jp

Bernard Swynghedauw
Centre de Recherches Cardiovasculaires de l'INSERM, Hôpital Lariboisière, 41 Bd de la Chapelle, 75475 Paris CEDEX, France, bernard.swynghedauw@larib.inserm.fr

Yasuharu Takaku
Center for Information Biology and DDBJ, National Institute of Genetics, Mishima, 411-8540 Shizuoka, Japan, ytakaku@lab.nig.ac.jp

Macarena Toll-Riera
Research Unit on Biomedical Informatics, Institut Municipal d'Investigació Mèdica, Universitat Pompeu Fabra, Barcelona, Spain, mtoll@imim.es

Feng Yu
School of Mathematics, University of Bristol, University Walk, Bristol BS8 1TW, UK, feng.yu@bristol.ac.uk

Part I
Modelization of Evolution

Chapter 1
Rate of Adaptation of Large Populations

Feng Yu and Alison Etheridge

Abstract We consider the accumulation of beneficial and deleterious mutations in asexual populations. The rate of adaptation is affected by the total mutation rate, proportion of beneficial mutations, and population size N. We describe two models: a strong selection model for large population size and a weak selection model for moderate population size. For the strong selection model, we give an argument to show that regardless of mutation rates, as long as the proportion of beneficial mutations is strictly positive, the adaptation rate is at least $\mathcal{O}(\log^{1-\delta} N)$, if the population size is sufficiently large. For the weak selection model, we use a Girsanov transform to perform some preliminary calculations that shed new light on the adaptation rate.

1.1 Background and Introduction

How does natural selection affect the rate of adaptation (defined to be the average speed at which mean fitness of the population increases)? To be more specific, let us assume we have a finite population of individuals, labelled $i = 1, \ldots, N$, each accumulating μ mutations per generation on average (each mutation is assumed to occur at a difference locus). A mutation is beneficial with probability q and deleterious with probability $1 - q$, In this work, we model all mutations to have the same absolute "fitness effects" σ, but their sign may be positive or negative depending on whether the mutation is beneficial or deleterious. We denote the selection coefficient σ, instead of s as in common in the biology literature, because we want to reserve s for a time variable that we use for calculations later on. Refer

---

F. Yu
School of Mathematics, University of Bristol, University Walk, Bristol BS8 1TW, UK
feng.yu@bristol.ac.uk

A. Etheridge
Department of Statistics, University of Oxford, 1 South Parks Road, Oxford OX1 3TG, UK
etheridg@stats.ox.ac.uk

to the description of model A in Sect. 1.2.1 and model B in Sect. 1.2.2 for the exact meaning of fitness effects $\pm\sigma$. A more sophisticated model should consider mutations that have a distribution of fitness effects, e.g. an independent exponentially distributed selective advantage associated with each new beneficial mutation as proposed by Gillespie (1991). Recent numerical work by Hegreness et al. (2006), however, suggests that in models where beneficial mutations have a distribution of fitness advantages, dynamics of adaptation can be reasonably described by an equivalent model where all beneficial mutations confer the same fitness advantage.

Without the effects of selection, we expect the rate of adaptation to be $\mu(2q-1)$. If $q < 1/2$, then the rate of adaptation is in fact negative, meaning that the population is becoming less fit with time. Selection should of course increase the rate of adaptation, but by how much? By Fisher's fundamental theorem (Fisher 1930), the adaptation rate is roughly proportional to the variance of the fitness among individuals in the population. It is well known that selection reduces the variance of fitness, but by how much and through what mechanisms this reduction occurs is not well understood mathematically. This understanding is a necessary step towards understanding the advantage of sex and recombination. Specifically, selection causes linkage disequilibrium and thus reduces the variance of fitness, whereas recombination reduces the linkage disequilibrium and thus increases the variance. The two mechanisms combine to induce a "recombination/selection balance".

Understanding the mechanisms through which selection and mutation interact with each other to determine the variance of fitness among individuals in the population and its adaptation rate also contributes to answering the following separate but related question: Does there exist an upper limit to the rate of adaptation in asexual populations? This question was originally posed to us by Nick Barton and can be more easily described in the context of purely beneficial mutations, i.e. take $q = 1$ in the last paragraph. Let us first consider the simplest scenario where a single beneficial mutation arises in an otherwise neutral population and no further mutations occur until the fate of that mutant is known. This situation is well understood, both in terms of p_{fix}, the fixation probability of the mutation, and the length of time it takes this process to complete. The value of p_{fix} was settled by Haldane (1927), who showed that if the selection coefficient associated with the mutation is σ, then under these circumstances $p_{fix} \approx 2\sigma$. In this case p_{fix} is almost independent of the population size, N.

If the mutation does fix, the process whereby it increases in frequency from $1/N$ to 1 is known as a *selective sweep*. The duration of a selective sweep is $\mathcal{O}(\log(\sigma N)/\sigma)$ generations. In particular, the length of a selective sweep is long if the population size N is large. If one assumes that the mutation rate per individual per generation is μ, then the overall mutation rate will be proportional to population size and we see that for large populations the assumption that no new mutation will arise during the time course of the sweep breaks down. Instead one expects multiple overlapping sweeps. Because there is no recombination, of all the mutations present in the population, only those on one lucky individual who eventually becomes the common ancestor of all individuals in the population will fix. Gerrish and Lenski (1998) called this effect *clonal interference*: clones that carry different beneficial mutations compete with each other and interfere with one another's growth in the population.

If the process has run long enough and is in some sort of stationarity, we can assume p_{fix} to be the same for all mutations, all of which are by assumption beneficial with fitness effect $+\sigma$. Furthermore, all mutations must eventually either fix or become extinct in finite time. Then the rate of adaptation is proportional to $\mu N p_{fix}$, where μN is the total number of mutations that occur to all individuals in the population in a single generation. If p_{fix} is independent of population size, then we expect an adaptation rate of $\mathcal{O}(N)$. But as the population size N becomes larger, the total number of mutations extant in the population at any particular time also grows, and thus the clonal interference will be more severe, reducing p_{fix}. This leads to the following question: If one does not limit the number of simultaneous selective sweeps, what is p_{fix}, or equivalently, what is the rate of adaptation? As $N \to \infty$, is the rate of adaptation finite or does it increase without bound? Barton and Coe (2007) suggest that there is an asymptotic limit to the rate of adaptation. Other authors (e.g. Rouzine et al. 2003; Wilke 2004; Desai and Fisher 2007) argue that no such limit exists.

Previous work on this question has adopted two general approaches: (1) calculate the fixation probability p_{fix} directly and (2) study the distribution of fitness of all individuals in the population and ask how this distribution evolves with time. The first approach was used in Gerrish and Lenski (1998), Wilke (2004) and Barton and Coe (2007), amongst others. This approach does not seem to be easy to adapt into a rigorous mathematical argument. Instead, we follow the second approach, i.e. to consider the distribution of fitnesses in the population, as in Rouzine et al. (2003, 2007) and Desai and Fisher (2007).

We will follow Rouzine et al. (2003) in taking fitness effects to be additive and all selection coefficients to be equal; thus, an individual's fitness can be characterised by the *net* number of beneficial mutations which it carries (which may be negative). Writing P_k for the proportion of individuals with fitness equivalent to k beneficial mutations, $\{P_k\}_{k \in \mathbb{Z}}$ forms a type of *travelling wave* whose shape remains basically unchanged over time. The position of the wave moves to the left or the right on the fitness axis, depending on whether the adaptation rate is positive or negative. The shape of the wave fluctuates stochastically even after a long time, and the fluctuations are larger if the population size is smaller. So the approximation of this *stochastic* wave by a *fixed* shape is most accurate when the population size is large. The approach is to treat the "bulk" of the wave as deterministic but the leading edge (i.e. the size of the fittest class) as stochastic. They found that the rate of adaptation, which is equal to the speed of the travelling wave, asymptotically depends logarithmically on population size N, which is consistent with results of in vitro studies of a type of RNA virus in Novella et al. (1995, 1999). Rouzine et al. (2007) offers an improved treatment of the stochastic edge.

Desai and Fisher (2007) followed a similar but slightly different approach. Just as in Rouzine et al. (2003), they studied the adaptation rate by studying how the sizes of different fitness classes change with time, and they also divided classes into deterministic and stochastic regimes. But instead of studying the wave speed directly, they examined the variance of fitness and studied how much variance can be maintained by mutation while the population is acted on by selection.

We have so far concentrated on beneficial mutations. In reality, most mutations are either neutral or deleterious. In particular, if all mutations in an asexual population were deleterious, the population would irreversibly accumulate deleterious mutations, a process known as Muller's ratchet. The first rigorous analysis of this phenomenon was due to Haigh (1978). Despite a very considerable body of work on Muller's ratchet, e.g. Higgs and Woodcock (1995), Stephan et al. (1993), Gordo and Charlesworth (2000) and Etheridge et al. (2007), a rigorous expression for the rate of decline in mean fitness of the population remains elusive. In this work, we are interested in both this rate of of decline in mean fitness and the question of whether a large population can overcome Muller's ratchet.

The conclusion we reach in Sect. 1.3 is the following: as long as the proportion of beneficial mutations is strictly positive, the rate of adaptation is roughly $\mathcal{O}(\log N)$ for large N, where N is the population size. Cuthbertson et al. (2007) established a rigorous lower bound of any fractional power of $\log N$ for the adaptation rate. This shows that no matter how small the proportion of beneficial mutations is, a large enough population size will yield a positive adaptation rate and Muller's ratchet is overcome. It also shows, in particular, that the rate of adaptation grows without bound as $N \to \infty$ in the all-mutations-beneficial case. This is consistent with the findings of Rouzine et al. (2003), Wilke (2004) and Desai and Fisher (2007). Furthermore, Barton and Coe (2007) argue that the adaptation rate in the case of unlinked loci and thus no linkage disequilibrium (i.e. links between loci are disrupted by very large recombination rates) is $\mathcal{O}(\log N)$. If this is true, then recombination can only increase the adaptation rate within the same order.

Throughout our work, we use continuous-time Moran models, where each individual has an exponentially distributed lifetime and mutations fall on each individual at a rate of μ. In the biological literature one would expect to see a Wright–Fisher model, where each individual has a lifetime of exactly one generation and fitness is often taken to be *multiplicative*. In the Moran model, individuals accumulate mutations during their lifetime, rather than at reproduction as in the Wright–Fisher model, but the average number of mutations accumulated by an individual during one unit of time in the Moran model can be made to correspond to the average number of mutations accumulated by an individual at each reproduction event in the Wright–Fisher model. Fitness in the context of Moran models is taken to be *additive* instead of multiplicative. For large populations, we expect that the results we obtain for our (much more mathematically tractable) Moran model will mirror those for the corresponding Wright–Fisher model.

Figure 1.1 shows a plot of the adaptation rate against log population size from simulation results of model A described in Sect. 1.2.1. We observe that for each set of parameters q, μ and σ, the rate of adaptation is roughly proportional to $\log N$ and small population sizes may result in negative adaptation rates. Furthermore, larger q results in a higher adaptation rate for fixed μ and σ. The upshot is that with μ and σ held constant, the smaller the proportion of beneficial mutations, the larger the population size required for Muller's ratchet to be overcome.

The argument we present in Sect. 1.3, as well as the argument used in Rouzine et al. (2003), works well only for large N, where the shape of the travelling wave can

1 Rate of Adaptation of Large Populations

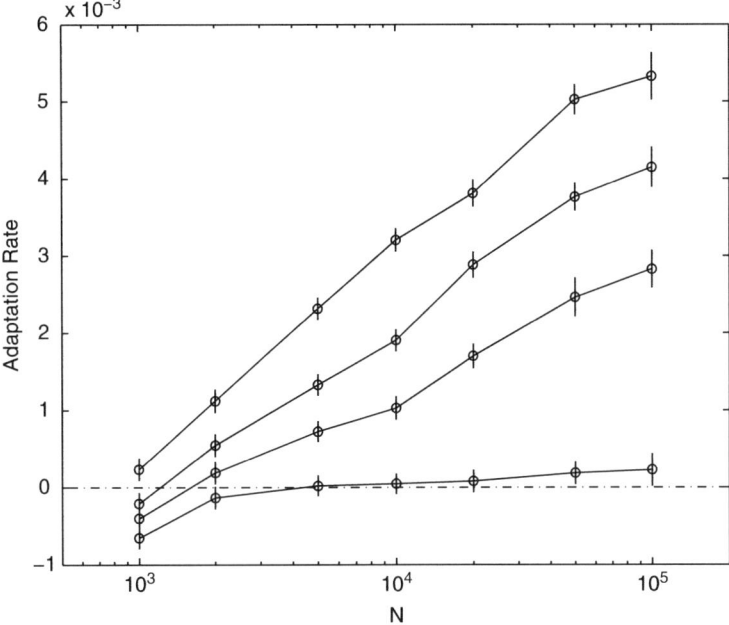

Fig. 1.1 Adaptation rate against population size, *from top to bottom*, for $q = 4, 2, 1$ and 0.2%, $\mu = 0.01$ and $\sigma = 0.01$. *Circles* represent actual data points and *vertical bars* represent 95% confidence intervals

be treated as deterministic. For moderate N, its shape actually fluctuates and an approximation by a fixed shape becomes unsatisfactory. See Fig. 1.2 for a comparison of the travelling wave for population sizes of $N = 1,000$ and $N = 100,000$; the shape for $N = 100,000$ is far more stable than that of $N = 1,000$. In the case of moderate population sizes, it may be more useful mathematically to rescale the mutation and selection parameters with population size to obtain a process where the proportions P_k of "individuals" in fitness class k can take any value in the interval $[0,1]$. Even though the limiting process is an infinite population limit, its behaviour still resembles a finite population process since by holding μN and σN constant and rescaling time by N, we maintain genetic drift as we take the limit $N \to \infty$. A limit we obtain this way is often called a *weak selection* model. This is what we do in Sect. 1.2.2. Because the sizes of essentially all fitness types are stochastic, studying the effects of mutation and selection on the adaptation rate requires a different approach and little mathematical analysis has so far been carried out. In Sect. 1.4, we show how to obtain the weak selection model from a neutral one with a Girsanov transform and then perform some calculations that shed new light on the adaptation rate. The aim of working with a weak selection model is still to understand the effect of mutation and selection on the adaptation rate, but this time for moderate population sizes.

The work is organised as follows. In Sect. 1.2, we formulate two models, a strong selection individual-based model and a weak selection diffusion model, and perform

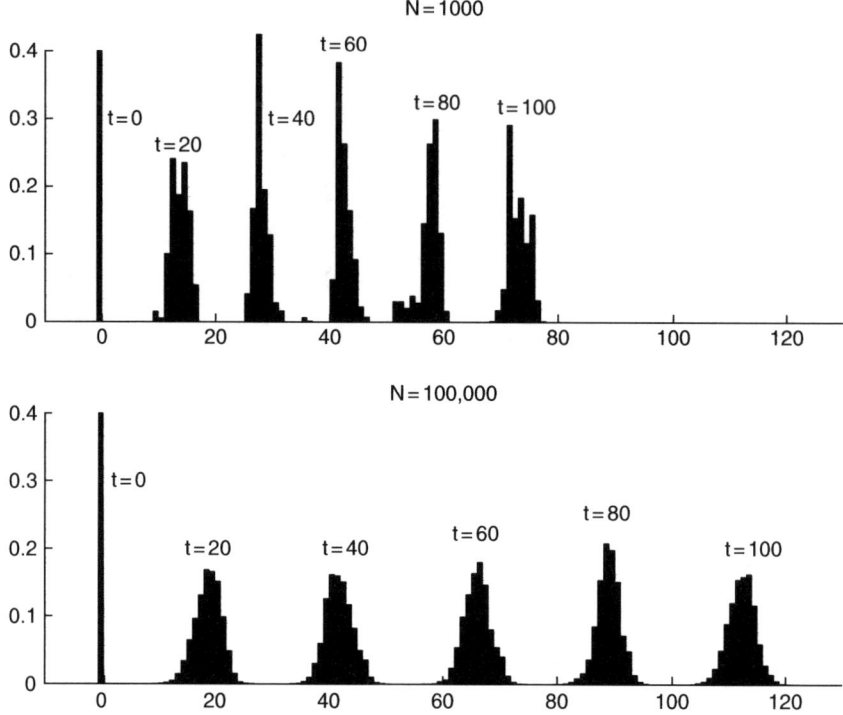

Fig. 1.2 The travelling wave for two populations sizes: $N = 1,000$ and $N = 100,000$. Both plots have parameters $q = 1$, $\mu = 0.02$ and $\sigma = 0.02$

some preliminary calculations on both models. In Sect. 1.3, we present a non-rigorous argument that leads to an asymptotic adaptation rate of roughly $\mathcal{O}(\log N)$ for the strong selection model. And finally in Sect. 1.4, we introduce the Girsanov transform to obtain the weak selection model, and show how it can yield useful information on the adaptation rate. Throughout this work, we focus on mathematical ideas that have the potential of being turned into rigorous arguments, rather than rigorous arguments themselves, which appear elsewhere.

1.2 Two Models

We first describe a finite population Moran model: the strong selection model. Then in Sect. 1.2.2 we describe a weak selection model in which mutation and selection coefficients are taken to be $\mathcal{O}(1/N)$. As $N \to \infty$, if time is measured in units of size N, the model converges to an infinite-dimensional diffusion.

We assume constant population size N. Let $X_i(t) \in \mathbb{Z}$, $i = 1, \ldots, N$, denote the *fitness type* of the ith individual, defined to be the number of beneficial mutations minus the number of deleterious mutations carried by the individual. We assume all mutations have the same absolute fitness effects; thus, the exact fitness of an individual is determined by its fitness type. Let $P_k(t)$ denote the proportion of individuals that have fitness k at time t, i.e.

$$P_k = \frac{1}{N} \sum_{i=1}^{N} \delta_{X_i=k}. \tag{1.1}$$

For $p \in \mathscr{P}^N(\mathbb{Z})$, the space of probability measures p on $\mathbb{Z}$ formed by N point masses each with weight $1/N$, define $p_k = p(\{k\})$ and

$$p_{[k,l]} = \sum_{i=k}^{l} p_i, \quad m_n(p) = \langle k^n, p \rangle = \sum_{k \in \mathbb{Z}} k^n p_k,$$
$$m(p) = m_1(p), \quad c_n(p) = \sum_{k \in \mathbb{Z}} (k - m(p))^n p_k.$$

In particular, $m(p)$ is the mean fitness of the population, and $c_2(p) = m_2(p) - m(p)^2$ is the second central moment of the population fitness, i.e. its variance.

The model of interest is one where each individual accumulates beneficial mutations at a Poisson rate $q\mu$ and deleterious mutations at rate $(1-q)\mu$. We assume a so-called infinitely-many-sites model where each mutation is assumed to be new and occur at a different locus on the genome.

1.2.1 Strong Selection Model

In the strong selection model, all individuals experience mutation, selection (which introduces a drift reflecting the differential reproductive success based on fitness) and genetic drift via resampling. This model, which we call *model A*, is described below:

1. *Mutation*: For each individual i a mutation event occurs at rate μ. With probability $1-q$, X_i changes to $X_i - 1$ and with probability q, X_i changes to $X_i + 1$.
2. *Selection*: For each pair of individuals (i,j), at rate $\frac{\sigma}{N}(X_i - X_j)^+$, individual i replaces individual j.
3. *Resampling*: For each pair of individuals (i,j), at rate $\frac{1}{2N}$, individual i replaces individual j.

This model has a time scale such that one unit of time corresponds roughly to one generation. Often one combines the resampling and selection into a single term. Each pair of individuals is involved in a reproduction event at some constant rate and the effect of selection is then that it is more likely to be the fitter individual that reproduces. In fact, this is the selection mechanism in model B, which we describe in Sect. 1.2.2. Since σ is typically rather small (one expects $N\sigma$ to be $\mathcal{O}(1)$) our formulation in model A is a very small perturbation of the model where resampling and selection are combined into a single mechanism, and does not affect our results.

The population $X_i(t)$ forms the empirical measure $P_k(t)$ by Eq. 1.1. We can also describe the mechanisms of model A in terms of the P_k's:

1. *Mutation*: For any $k \in \mathbb{Z}$, at rate $(1-q)\mu N P_k$, P_k decreases by $\frac{1}{N}$ and P_{k-1} increases by $\frac{1}{N}$; at rate $q\mu N P_k$, P_k decreases by $\frac{1}{N}$ and P_{k+1} increases by $\frac{1}{N}$.
2. *Selection*: For any pair of $k,l \in \mathbb{Z}$ such that $k > l$, at rate $\sigma(k-l)NP_k P_l$, P_k increases by $\frac{1}{N}$ and P_l decreases by $\frac{1}{N}$.
3. *Resampling*: For any pair of $k,l \in \mathbb{Z}$, at rate $\frac{N}{2}P_k P_l$, P_k increases by $\frac{1}{N}$ and P_l decreases by $\frac{1}{N}$.

We perform a so-called martingale decomposition for P_k in Eq. 1.2 below, i.e. we write P_k as a sum of deterministic drifts and martingales, where, roughly, the drift describes the average evolution of P_k and the martingales describe how much the evolution can differ from the average behaviour. For details of this type of calculation, see Ikeda and Watanabe (1981). For this work, we just point out that a pure jump Markov process that increases by $1/N$ at rate $r(t)$ decomposes into a drift term $\frac{1}{N}\int_0^t r(s)\,ds$ and a martingale term that has quadratic variation $\frac{1}{N^2}\int_0^t r(s)\,ds$.

$$P_k(t) = P_k(0) + \mu \int_0^t [q(P_{k-1}(s) - P_k(s)) + (1-q)(P_{k+1}(s) - P_k(s))]\,ds$$
$$+ \sigma \int_0^t \sum_{l \in \mathbb{Z}} (k-l) P_k(s) P_l(s)\,ds + M_k^{P,1}(t) + M_k^{P,2}(t)$$
$$= P_k(0) + \mu \int_0^t (qP_{k-1}(s) - P_k(s) + (1-q)P_{k+1}(s))\,ds$$
$$+ \sigma \int_0^t \sum_{l \in \mathbb{Z}} (k-l) P_k(s) P_l(s)\,ds + M_k^{P,1}(t) + M_k^{P,2}(t), \quad (1.2)$$

where $M_k^{P,1}$ and $M_k^{P,2}$ are orthogonal martingales, the first arising from the (compensated) mutation mechanism and the second from the resampling and (compensated) selection mechanisms. Both have maximum jump size $1/N$ and their quadratic variations are all $\mathcal{O}(1/N)$, i.e. $[M_k^{P,1}](t) = \mathcal{O}(1/N)$ and $[M_k^{P,2}](t) = \mathcal{O}(1/N)$. For exact formulae for these quadratic variations, see Cuthbertson et al. (2007).

We can write the martingale decomposition of the mean $m(P(t)) = \sum_k k P_k(t)$ as follows:

$$m(P(t)) = m(P(0)) + M^{P,m}(t) + \int_0^t \sigma \sum_{k,l \in \mathbb{Z}} k(k-l) P_k(s) P_l(s)$$
$$+ \mu \sum_{k \in \mathbb{Z}} k(qP_{k-1}(s) - P_k(s) + (1-q)P_{k+1}(s))\,ds \quad (1.3)$$
$$= m(P(0)) + M^{P,m}(t) + \int_0^t \sigma [m_2(P(s)) - m(P(s))^2]$$
$$+ \mu \sum_{k \in \mathbb{Z}} [q(k+1) - 1 + (1-q)(k-1)] P_k(s)\,ds$$
$$= m(P(0)) + M^{P,m}(t) + \int_0^t \sigma [c_2(P(s)) + \mu(2q-1)]\,ds,$$

where $M^{P,m}$ is a martingale whose quadratic variation is $\mathcal{O}(1/N)$. In differential notations, we write

$$dm(P) = [\mu(2q-1) + \sigma c_2(P)]\,dt + dM^{P,m}. \tag{1.4}$$

1.2.2 Weak Selection Model

For the weak selection model, we let $\tilde{\mu}$ and $\tilde{\sigma}$ denote the mutation and selection coefficients, respectively, and define

$$\tilde{\mu} = N\mu, \quad \tilde{\sigma} = N\sigma,$$

where μ and σ are the mutation and selection coefficients in the strong selection model defined in Sect. 1.2.1. To obtain the mechanisms of the weak selection model, we hold $\tilde{\mu}$ and $\tilde{\sigma}$ constant (i.e. scale μ and σ down by N), combine the selection and resampling mechanism into a single reproduction mechanism, and then speed up time by N, to obtain the following Moran particle model with constant population size N, which we call *model B*:

1. *Mutation*: For each individual i a mutation event occurs at rate $\tilde{\mu}$. With probability $1-q$, X_i changes to $X_i - 1$ and with probability q, X_i changes to $X_i + 1$.
2. *Reproduction*: For a pair of individuals (i,j), a reproduction event occurs at rate 1. With probability $\frac{1}{2}(1 + \frac{\tilde{\sigma}}{N}(X_i - X_j))$, individual i replaces individual j; with probability $\frac{1}{2}(1 - \frac{\tilde{\sigma}}{N}(X_i - X_j))$, individual j replaces individual i.

In one unit of time, each individual experiences roughly N reproduction events; thus, one unit of time corresponds to N generations in this model, and consequently $\tilde{\mu}$ and $\tilde{\sigma}$ correspond with $N\mu$ and $N\sigma$, respectively. Of course, the definition of this reproduction mechanism runs into trouble if any two individuals i and j have fitness difference larger than $N/\tilde{\sigma}$, i.e. if $|X_i - X_j| > N/\tilde{\sigma}$, in which case one of the two probabilities in the reproduction mechanism becomes negative and the other becomes larger than 1. The probability of this happening will converge to 0 as $N \to \infty$ and we ignore this possibility in the calculations below.

We use $P_k^{(N)}$ to denote the proportion of individuals of fitness type k, to emphasise the fact that this is an empirical measure formed by N individuals each with mass $1/N$, and that the process is a jump process with jump size $1/N$. We will eventually take an infinite population limit $N \to \infty$ and obtain a process with continuous paths. In terms of the proportions $P_k^{(N)}$'s, model B can be described as follows:

1. *Mutation*: For any $k \in \mathbb{Z}$, at rate $(1-q)\tilde{\mu}NP_k^{(N)}$, $P_k^{(N)}$ decreases by $\frac{1}{N}$ and $P_{k-1}^{(N)}$ increases by $\frac{1}{N}$; at rate $q\tilde{\mu}NP_k^{(N)}$, $P_k^{(N)}$ decreases by $\frac{1}{N}$ and $P_{k+1}^{(N)}$ increases by $\frac{1}{N}$.
2. *Reproduction*: For any pair of $k,l \in \mathbb{Z}$, at rate $\frac{N^2}{2}(1 + \frac{\tilde{\sigma}}{N}(k-l))P_k^{(N)}P_l^{(N)}$, $P_k^{(N)}$ increases by $\frac{1}{N}$ and $P_l^{(N)}$ decreases by $\frac{1}{N}$; at rate $\frac{N^2}{2}(1 - \frac{\tilde{\sigma}}{N}(k-l))P_k^{(N)}P_l^{(N)}$, $P_l^{(N)}$ increases by $\frac{1}{N}$ and $P_k^{(N)}$ decreases by $\frac{1}{N}$.

Just as for model A, we can perform the martingale decomposition to obtain

$$P_k^{(N)}(t) = P_k^{(N)}(0) + \tilde{\mu} \int_0^t \left(q P_{k-1}^{(N)}(s) - P_k^{(N)}(s) + (1-q) P_{k+1}^{(N)}(s) \right) ds$$

$$+ \tilde{\sigma} \int_0^t \sum_{l \in \mathbb{Z}} (k-l) P_k^{(N)}(s) P_l^{(N)}(s) \, ds + M_k^{(N),P,1}(t) + M_k^{(N),P,2}(t),$$

$$= P_k^{(N)}(0) + \tilde{\mu} \int_0^t \left(q P_{k-1}^{(N)}(s) - P_k^{(N)}(s) + (1-q) P_{k+1}^{(N)}(s) \right) ds$$

$$+ \tilde{\sigma} \int_0^t \left(k - m\left(P^{(N)}(s)\right) \right) P_k^{(N)}(s) \, ds + M_k^{(N),P,1}(t) + M_k^{(N),P,2}(t),$$

where $M_k^{(N),P,1}$ and $M_k^{(N),P,2}$ are orthogonal martingales, the first arising from the (compensated) mutation mechanism and the second from the reproduction mechanism. Because we have sped up time by a factor of N, the quadratic variation of $M^{(N),P,2}$ (which we recall arises from the resampling) is no longer of $\mathcal{O}(1/N)$:

$$\left[M_k^{(N),P,1} \right](t) = \mathcal{O}(1/N),$$

$$\left[M_k^{(N),P,1}, M_l^{(N),P,1} \right](t) = 0 \text{ if } |k-l| \geq 2,$$

$$\left[M_k^{(N),P,2} \right](t) = \int_0^t \sum_{l \in \mathbb{Z}} P_k^{(N)}(s) P_l^{(N)}(s) \, ds$$

$$= \int_0^t P_k^{(N)}(s)(1 - P_k^{(N)}(s)) \, ds,$$

$$\left[M_k^{(N),P,2}, M_l^{(N),P,2} \right](t) = - \int_0^t P_k^{(N)}(s) P_l^{(N)}(s) \, ds \text{ if } k \neq l.$$

Because $M^{(N),P,2}$ is $\mathcal{O}(1)$, it persists even after taking the limit $N \to \infty$, while $M^{(N),P,1}$ goes away since it is $\mathcal{O}(1/N)$.

We take the limit $N \to \infty$ to obtain a process $P_k(t)$ that is continuous for each $k \in \mathbb{Z}$ and $t \geq 0$ and satisfies the martingale problem

$$P_k(t) = P_k(0) + \tilde{\mu} \int_0^t (q P_{k-1}(s) - P_k(s) + (1-q) P_{k+1}(s)) \, ds$$

$$+ \tilde{\sigma} \int_0^t (k - m(P(s))) P_k(s) \, ds + M_k^P(t), \quad (1.5)$$

where M_k^P are martingales with quadratic variation process

$$\left[M_k^P, M_{k'}^P \right](t) = \int_0^t P_k(s)(\delta_{kk'} - P_{k'}(s)) \, ds, \quad (1.6)$$

where $\delta_{kk'} = 1$ if $k = k'$ and $\delta_{kk'} = 0$ otherwise.

1 Rate of Adaptation of Large Populations

The existence and uniqueness of the solution of this martingale problem is non-trivial, but it will appear elsewhere and we do not go into detail here. The martingales M_k^P can be represented in terms of Brownian motions in the following way:

$$M_k^P(t) = \int_0^t \sum_{l \in \mathbb{Z}} \sqrt{P_k(s)P_l(s)}\, dW_{kl}(s), \tag{1.7}$$

where $\{W_{kl} : k,l \in \mathbb{Z}, k > l\}$ are independent Brownian motions and we define $W_{kl} = -W_{lk}$ for $k,l \in \mathbb{Z}$ and $k < l$ and $W_{kk} \equiv 0$ for $k \in \mathbb{Z}$. We now verify that M_k^P has the quadratic variation process specified by Eq. 1.6. We write $P_k = P_k(s)$ if this causes no confusion. In differential notation,

$$d[M_k^P, M_{k'}^P] = \sum_{l,l'} \sqrt{P_k P_l P_{k'} P_{l'}}\, d[W_{kl}, W_{k'l'}]$$

for $k \neq l$. Since $k \neq l$, $d[W_{kl}, W_{k'l'}] = -dt$ if $k = l'$ and $l = k'$ and $d[W_{kl}, W_{k'l'}] = 0$ otherwise. Therefore,

$$d[M_k^P, M_{k'}^P] = -\sqrt{P_k P_l P_l P_k}\, dt = -P_k P_l\, dt.$$

Similarly,

$$d[M_k^P] = \sum_{l,l'} \sqrt{P_k^2 P_{k'} P_{l'}}\, d[W_{kl}, W_{kl'}]$$

$$= \sum_l \sqrt{P_k^2 P_l^2}\, d[W_{kl}, W_{kl}] + \sum_{l,l': l \neq l'} \sqrt{P_k^2 P_l P_{l'}}\, d[W_{kl}, W_{kl'}].$$

Since $d[W_{kl}, W_{kl}] = dt$ if $l \neq k$, and if $l \neq l'$ then $d[W_{kl}, W_{kl'}] = 0$, we have

$$d[M_k^P] = \sum_{l \neq k} P_k P_l\, dt = P_k(1 - P_k)\, dt.$$

Thus, we can write the solution to the martingale problem (Eq. 1.5) as the solution to the following infinite system of stochastic differential equations (SDEs),

$$dP_k = [\tilde{\mu}(qP_{k-1} - P_k + (1-q)P_{k+1}) + \tilde{\sigma}(k - m(P))P_k]\, dt + \sum_{l \in \mathbb{Z}} \sqrt{P_k P_l}\, dW_{kl}, \tag{1.8}$$

and then associate an infinite-dimensional operator $\mathcal{A}_{\tilde{\sigma},q,\tilde{\mu}}$ with this system:

$$\mathcal{A}_{\tilde{\sigma},q,\tilde{\mu}} f(p) = \sum_k [\tilde{\mu}(qp_{k-1} - p_k + (1-q)p_{k+1}) + \tilde{\sigma}(k - m(p))p_k] \frac{\partial f}{\partial p_k}$$

$$+ \frac{1}{2} \sum_{k,l} p_k(\delta_{kl} - p_l) \frac{\partial^2 f}{\partial p_k \partial p_l}. \tag{1.9}$$

The operator $\mathcal{A}_{\tilde{\sigma},q,\tilde{\mu}}f(p)$ is usually called the generator of the stochastic process and its domain can be taken to be $C^2(\mathscr{P}(\mathbb{Z}))$, the space of twice continuously differentiable functions on $\mathscr{P}(\mathbb{Z})$, the space of probability measures on $\mathbb{Z}$. For convenience, we will refer to the stochastic process that solves Eq. 1.8 as the process associated with the generator $\mathcal{A}_{\tilde{\sigma},q,\tilde{\mu}}$.

If we set $\tilde{\sigma} = 0$, then the process associated with $\mathcal{A}_{0,q,\tilde{\mu}}$ solves the SDE

$$dP_k = [\tilde{\mu}(qP_{k-1} - P_k + (1-q)P_{k+1})]\,dt + \sum_{l \in \mathbb{Z}} \sqrt{P_k P_l}\, dW_{kl},$$

and can be thought of as the *neutral* process that has only the mutation (both beneficial and deleterious) and neutral resampling mechanisms. Because the resampling mechanism is neutral, many techniques, e.g. the look-down construction (Donnelly and Kurtz 1999), can be brought to bear on this process and consequently many quantities can be calculated for the neutral process. The main technique we use to study the selected process $\mathcal{A}_{\tilde{\sigma},q,\tilde{\mu}}$ is to write it as a Girsanov transform of the neutral process $\mathcal{A}_{0,q,\tilde{\mu}}$. We will discuss this technique in more detail in Sect. 1.4.

1.3 Strong Selection: Asymptotic Adaptation Rate

In this section, we take large population size and focus on the strong selection model A that we described in Sect. 1.2.1. We give a non-rigorous argument that leads to an asymptotic adaptation rate of roughly $\mathcal{O}(\log N)$, as long as q is strictly positive and regardless of the selection and mutation parameters.

The basic idea of this approach is to treat the "bulk" of the travelling wave (where $P_k = \mathcal{O}(1)$) as deterministic and only treat the front and tail of this wave (where $P_k \ll 1$) as stochastic. This is similar to the approach taken in Rouzine et al. (2003), where an approximate expression for the shape of wave is first derived using a deterministic equation. There is an infinite family of solutions to this deterministic equation, parameterised by the wave speed. To determine the correct wave speed for a given population size, stochasticity at the front of the wave must be taken into account. In our work, we derive a set of moment equations and show that the selection mechanism dictates that the form of the wave is approximately Gaussian. We then calculate the speed of this wave in two ways, one using the bulk, the other using the front. These two speeds must be the same, which gives us a constraint that determines our approximation to the adaptation rate.

With the martingale term $M^{P,1}$ and $M^{P,2}$ of $\mathcal{O}(1/\sqrt{N})$, the effect of noise on P_k can be considered to be quite small if P is much larger than $1/N$. For P in this range, we have from Eq. 1.2,

$$\begin{aligned}dP_k &\approx \left[\mu(qP_{k-1} - P_k + (1-q)P_{k+1}) + \sigma \sum_{l \in \mathbb{Z}}(k-l)P_k P_l\right]dt \\ &= [\mu(qP_{k-1} - P_k + (1-q)P_{k+1}) + \sigma(k - m(P))P_k]\,dt.\end{aligned} \qquad (1.10)$$

This is similar to Eq. 2 in Rouzine et al. (2003), except that they treated discrete time. With P_k evolving according to Eq. 1.10, we can write the evolution of the mean fitness $m(P)$ in the following way:

$$dm(P) = (\mu(2q-1) + \sigma c_2(P))\,dt, \qquad (1.11)$$

which is simply Eq. 1.4 on dropping the martingale term $M^{P,m}$. We can use the above two equations to calculate the evolution of $c_n(p) = \sum_k (k-m(p))^n p_k$:

$$\begin{aligned}
\frac{dc_n(P)}{dt} &= \sum_k (k-m(P))^n [\mu(qP_{k-1} - P_k + (1-q)P_{k+1}) + \sigma(k-m(P))P_k] \\
&\quad - \sum_k n(k-m(P))^{n-1}(\mu(2q-1) + \sigma c_2(P))P_k \\
&= \mu \sum_k (k-m(P))^n [qP_{k-1} - P_k + (1-q)P_{k+1}] - n(k-m(P))^{n-1}(2q-1)P_k \\
&\quad + \sigma(c_{n+1} - nc_{n-1}c_2) \\
&= \mu \sum_k (k-m(P))^{n-1} [q(k+1-m(P)) - (k-m(P)) \\
&\quad + (1-q)(k-1-m(P)) - n(2q-1)]P_k + \sigma(c_{n+1} - nc_{n-1}c_2).
\end{aligned}$$

Thus,

$$\frac{dc_n(P)}{dt} = \sigma(c_{n+1}(P) - nc_{n-1}(P)c_2(P)). \qquad (1.12)$$

Notice that Eq. 1.12 does not depend on the mutation coefficient μ, and thus the mutation mechanism does not affect the overall shape of the wave. Since the martingale term $M^{P,m}$ that we dropped in Eq. 1.4 to obtain Eq. 1.11 is small compared with P_k where P_k is significantly larger than $1/N$, we can assume the shape of the wave, as described by the central moments c_n, evolves approximately deterministically according to Eq. 1.12. Setting the right-hand side of Eq. 1.12 to zero for all $n \geq 2$, we see that the shape of P is approximately Gaussian, although Eq. 1.12 admits any Gaussian distribution as its solution regardless of its variance. In other words, Eq. 1.12 has an infinite family of solutions, each a Gaussian distribution, and the system itself yields no information on the exact value of the variance.

It is the behaviour at the front and tail of the wave that determines the value of this variance (and hence the wave speed). Suppose then that P is approximately Gaussian with mean fitness $m(P)$ and variance b^2. The population size N is large but finite so the "front" of the wave is approximately where the level of P reaches $1/N$ (i.e. one individual). This level is roughly at $K + m(P)$ where K satisfies

$$\frac{1}{2\pi b^2} e^{-K^2/2b^2} = \frac{1}{N},$$

which implies that

$$K \approx b\sqrt{2\log N}. \qquad (1.13)$$

We assume that a single individual is born at site $K+m(P)$ at time $t=0$ and try to find out how long it takes for an individual to be born at site $K+1+m(P)$. This is how long it takes the wave to advance by one. Let $Z(t)$ be the number of individuals of the fittest type, that is the size of the population at site $K+m(P)$, then Eq. 1.10 says that until a beneficial mutation falls on site $K+m(P)$, $Z(t)$ increases exponentially at rate $\sigma K - \mu$, ignoring beneficial mutations occurring to type $K-1+m(P)$, i.e.

$$Z(t) \approx e^{(\sigma K - \mu)t}. \tag{1.14}$$

While the number of individuals at site $K+m(P)$ grows, all of them are accumulating beneficial and deleterious mutations at rates $q\mu$ and $(1-q)\mu$, respectively. As soon as one of these individuals accumulates a beneficial mutation, the "front" of the wave will have advanced by one. The probability that no beneficial mutation occurs to any individuals with fitness $K+m(P)$ is

$$\exp\left(-q\mu \int_0^t Z(s)\,ds\right) = \exp\left(-\frac{q\mu}{\sigma K - \mu}\left(e^{(\sigma K - \mu)t} - 1\right)\right),$$

using the expression for $Z(t)$ in Eq. 1.14. Therefore, the average time when the front advances by one can be calculated to be

$$\frac{e^{q\mu/(\sigma K - \mu)}}{\sigma K - \mu} \int_{1/(\sigma K - \mu)}^{\infty} \frac{e^{-q\mu u}}{u}\,du.$$

Using the fact that for small ε, $\int_\varepsilon^\infty e^{-q\mu u}/u\,du = \log(1/\varepsilon) + \mathcal{O}(1)$, we deduce that the average time until the front advances by one is roughly $\frac{1}{\sigma K - \mu}\log(\sigma K - \mu)$ for large K, which gives a wave speed of $(\sigma K - \mu)/\log(\sigma K - \mu)$.

But by Eq. 1.11, the wave speed is $\mu(2q-1) + \sigma c_2(P) = \mu(2q-1) + \sigma b^2 \approx \mu(2q-1) + \sigma K^2/(2\log N)$, using Eq. 1.13. This leads to the following consistency condition:

$$\frac{\sigma K - \mu}{\log(\sigma K - \mu)} = \mu(2q-1) + \frac{\sigma K^2}{2\log N}.$$

For large K, this approximately reduces to

$$K\log(\sigma K) = 2\log N.$$

This is a transcendental equation with no closed-form solution. If $K = \log N$, then the left-hand side is greater than the right-hand side; if, on the other hand, $K = \log^{1-\delta} N$ for any small positive δ, then left-hand side is less than the right-hand side. So K is between $\log N$ and any fractional power of $\log N$, i.e. a bit smaller than $\mathcal{O}(\log N)$, which implies a wave speed, that is a rate of adaptation, of order between $\log N$ and any fractional power of $\log N$.

In Cuthbertson et al. (2007), we in fact established a rigorous asymptotic lower bound of $\log^{1-\delta} N$ (where δ is any small positive number) for the adaptation rate, as long as q is strictly positive and regardless of the selection and mutation parameters. The proof uses some of the ideas outlined in this section.

1.4 Weak Selection: Girsanov Calculations

Now we turn to the weak selection model of Sect. 1.2.2. We would like to study the adaptation rate of the *selected* process associated with generator $\mathcal{A}_{\tilde{\sigma},q,\tilde{\mu}}$ (defined in Eq. 1.9) with $\tilde{\sigma} \neq 0$. We can get a set of moment equations similar to Eq. 1.12, which is not closed and is therefore difficult to study. The *neutral* process associated with $\mathcal{A}_{0,q,\tilde{\mu}}$, on the other hand, is much easier to study. The central idea of our approach is to find a relationship between the probability measure associated with the selected process and that associated with the neutral process. The technique we use is the Girsanov transform.

1.4.1 The Girsanov Transform

The Girsanov theorem tells us how the stochastic process changes if we change the underlying probability measure. We review this briefly for the simplest case of adding drift to a one-dimensional standard Brownian motion. Suppose $W(t)$ is a Brownian motion under the probability measure $\mathbb{P}$. Let $\{\mathcal{F}_t\}_{t \geq 0}$ be the corresponding filtration. Let $a(t)$ be an $\mathcal{F}_t$-adapted process, then $Z(t) = \exp\{\int_0^t a(s)\, dB(s) - \frac{1}{2}\int_0^t a(s)^2\, ds\}$ is a local martingale. If $a(t)$ satisfies the Novikov condition, i.e. $E[\exp\{\int_0^t a(s)^2\, ds\}] < \infty$, then $Z(t)$ is actually a martingale and we can define a probability $\mathbb{Q}$ by

$$\left.\frac{d\mathbb{Q}}{d\mathbb{P}}\right|_{\mathcal{F}_t} = Z(t).$$

By the Girsanov theorem (see Karatzas and Shreve 1991), the process

$$X(t) = X(0) - \int_0^t a(s)\, ds + B(t)$$

is a Brownian motion under $\mathbb{Q}$. This gives us a way of relating the distribution of the standard Brownian motion and that of a Brownian motion with drift.

The Girsanov theorem works equally well when the Brownian motion described above is replaced with more general continuous martingales or the dimension is larger than 1 but finite. For the infinite-dimensional setting, we can use Dawson's Girsanov theorem (see Dawson 1991 and Theorem 7.9 of Etheridge 2000) developed for Dawson–Watanabe superprocesses. We first recall the *neutral* model that we obtained in Sect. 1.2.2 as the infinite population limit of processes described by model B. Setting $\tilde{\sigma} = 0$ in Eq. 1.8, the neutral process satisfies the following infinite system of SDEs:

$$dP_k = [\tilde{\mu}(qP_{k-1} - P_k + (1-q)P_{k+1})]\,dt + dM_k, \tag{1.15}$$

where we define the martingale

$$M_k(t) = \int_0^t \sum_{l \in \mathbb{Z}} \sqrt{P_k(s)P_l(s)}\,dW_{kl}(s).$$

We associate this process with its generator $\mathcal{A}_{0,q,\tilde{\mu}}$, with $\mathcal{A}_{\tilde{\sigma},q,\tilde{\mu}}$ defined in Eq. 1.9 for any $\tilde{\sigma}$. Let $\mathbb{P}_0$ denote the law of $\{P_k, k \in \mathbb{Z}\}$ evolving according to the neutral generator $\mathcal{A}_{0,q,\tilde{\mu}}$, then by definition, M_k is a martingale under $\mathbb{P}_0$. We define the function $a_n(p)$

$$a_n(p) = \tilde{\sigma}\left(n - \sum_k k p_k\right) = \tilde{\sigma}(n - m(p))$$

and Z to be a process that satisfies the following SDE:

$$dZ = Z \sum_n a_n\,dM_n = \tilde{\sigma} Z \sum_n (n - m(p))\,dM_n. \tag{1.16}$$

In particular, if the Novikov condition is satisfied, i.e. if

$$\mathbb{E}^{\mathbb{P}_0}\left[\exp\left(\tilde{\sigma}^2 \int_0^T c_2(P(s))\,ds\right)\right] < \infty,$$

then Z is a martingale up to time T under $\mathbb{P}_0$. The Novikov condition does hold for the process $\mathcal{A}_{0,q,\tilde{\mu}}$ under the stationary measure of the process centred about its mean, but we skip the details, which will appear elsewhere. We define

$$\tilde{M}_k(t) = M_k(t) - \tilde{\sigma}\int_0^t \sum_n a_n(s)\,d[M_k, M_n](s) \tag{1.17}$$

$$= M_k(t) - \tilde{\sigma}\int_0^t \sum_n (n - m(P(s)))(\delta_{kn} - P_k(s))P_n(s)\,ds$$

$$= M_k(t) - \tilde{\sigma}\int_0^t \left[(k - m(P(s)))P_k(s) - \sum_n (n - m(P(s)))P_k(s)P_n(s)\right]ds$$

$$= M_k(t) - \tilde{\sigma}\int_0^t (k - m(P(s)))P_k(s)\,ds,$$

where we use Eq. 1.6 in the second line. By Dawson's Girsanov theorem, $\tilde{M}_k$ is a $\mathbb{P}_{\tilde{\sigma}}$-martingale, where

$$\left.\frac{d\mathbb{P}_{\tilde{\sigma}}}{d\mathbb{P}_0}\right|_{\mathscr{F}_t} = Z(t)$$

and

$$[M_k, M_l](t) = [\tilde{M}_k, \tilde{M}_l](t).$$

1 Rate of Adaptation of Large Populations

Although they are martingales under different probability measures, the $\tilde{M}_k$'s and M_k's have the same cross-variation structures; thus as in Eq. 1.7 for M_k, we can define $\{\tilde{W}_{kl} : k, l \in \mathbb{Z}, k > l\}$ to be independent Brownian motions, $\tilde{W}_{kl} = -\tilde{W}_{lk}$ for $k, l \in \mathbb{Z}$ and $k < l$, and $\tilde{W}_{kk} \equiv 0$ for $k \in \mathbb{Z}$. Then

$$\tilde{M}_k^P(t) = \int_0^t \sum_{l \in \mathbb{Z}} \sqrt{P_k(s) P_l(s)} \, d\tilde{W}_{kl}(s) \tag{1.18}$$

are martingales under $\mathbb{P}_{\tilde{\sigma}}$. From the definition of $\tilde{M}$ in Eq. 1.17, we see that M satisfies

$$dM_k = d\tilde{M}_k(t) + \tilde{\sigma}(k - m(P(t))) P_k(t) \, dt. \tag{1.19}$$

From Eq. 1.15, we easily obtain $dM_k = dP_k - [\tilde{\mu}(qP_{k-1} - P_k + (1-q)P_{k+1})] \, dt$. We plug this into the left-hand side of Eq. 1.19 and Eq. 1.18 into its right-hand side to obtain

$$dP_k = [\tilde{\mu}(qP_{k-1} - P_k + (1-q)P_{k+1}) + \tilde{\sigma}(k - m(P))P_k] \, dt + \sum_{l \in \mathbb{Z}} \sqrt{P_k P_l} \, d\tilde{W}_{kl}.$$

We see this is exactly the SDE in Eq. 1.8 with generator $\mathcal{A}_{\tilde{\sigma},q,\tilde{\mu}}$, except the Brownian motions W_{kl} are replaced with another family of Brownian motions $\tilde{W}_{kl}$.

To summarise, W_{kl}'s are Brownian motions under $\mathbb{P}_0$, which is the law for the process P associated with $\mathcal{A}_{0,q,\tilde{\mu}}$, while $\tilde{W}_{kl}$'s are Brownian motions under $\mathbb{P}_{\tilde{\sigma}}$, the law for P associated with $\mathcal{A}_{\tilde{\sigma},q,\tilde{\mu}}$. The Radon–Nikodym derivative $(d\mathbb{P}_{\tilde{\sigma}}/d\mathbb{P}_0)|_{\mathcal{F}_t}$ is given by $Z(t)$, which is a martingale that satisfies Eq. 1.16 and can be written out explicitly:

$$Z(t) = \exp\left(\int_0^t \sum_k a_k(P(s)) \, dM_k(s) - \frac{1}{2} \int_0^t \sum_{k,l} a_k(P(s)) a_l(P(s)) \, d[M_k, M_l](s) \right).$$

We now find a more convenient expression for $Z(t)$. The first integral in the exponent is

$$\int_0^t \sum_k a_k(P(s)) \, dM_k(s) = \tilde{\sigma} \int_0^t \sum_{k,l} (k - m(P(s))) \sqrt{P_k(s) P_l(s)} \, dW_{kl}(s)$$

$$= \tilde{\sigma} \int_0^t \sum_k k \sum_l \sqrt{P_k(s) P_l(s)} \, dW_{kl}(s) = \tilde{\sigma} \sum_k k M_k(s),$$

where the second equality comes from the fact $\sum_{k,l} \sqrt{P_k(s) P_l(s)} \, dW_{kl}(s) = 0$, since $W_{kl} = -W_{lk}$. By Eq. 1.15,

$$\sum_k k M_k(t) = \sum_k k P_k(t) - \sum_k k P_k(0) - \int_0^t \sum_k k [\tilde{\mu}(q P_{k-1}(s) - P_k(s) + (1-q) P_{k+1}(s))] \, ds$$

$$= m(P(t)) - m(P(0)) - \tilde{\mu}(2q - 1)t.$$

The second integral in the exponent is

$$\int_0^t \sum_{k,l} a_k(P(s)) a_l(P(s))\, d[M_k, M_l](s)$$
$$= \tilde{\sigma}^2 \int_0^t \sum_{k,l} (k - m(P(s)))(l - m(P(s)))(\delta_{kl} - P_k(s)) P_l(s)\, ds$$
$$= \tilde{\sigma}^2 \int_0^t \sum_k (k - m(p(s)))^2 P_k(s)\, ds = \tilde{\sigma}^2 \int_0^t c_2(P(s))\, ds.$$

Therefore,

$$Z(t) = \exp\left(\tilde{\sigma}[m(P(t)) - m(P(0)) - \tilde{\mu}(2q-1)t] - \frac{\tilde{\sigma}^2}{2} \int_0^t c_2(P(s))\, ds\right),$$

where P solves the neutral SDE (Eq. 1.15).

1.4.2 Moments of the Neutral Process

For the calculations on the selected process (Eq. 1.8) that we perform in Sect. 1.4.3, we need to know some asymptotic moments of the neutral process (Eq. 1.15). We drop the notational dependence on p of the moments, m, c_2, c_3, etc. The mean of P obeys

$$dm = \tilde{\mu} \sum_k k[(qP_{k-1} - P_k + (1-q)P_{k+1})]\, dt + dM$$
$$= \tilde{\mu}(2q-1)\, dt + dM, \qquad (1.20)$$

where we define the martingale

$$M(t) = \sum_k k M_k(t),$$

which has quadratic variation

$$d[M] = \sum_{k,l,k',l'} k\sqrt{P_k P_l}\, k'\sqrt{P_{k'} P_{l'}}\, d[W_{kl}, W_{k'l'}]$$
$$= \sum_{k,l,k',l':k'=k,l'=l} kk'\sqrt{P_k P_l P_{k'} P_{l'}}\, dt - \sum_{k,l,k',l':k=l',k'=l} kk'\sqrt{P_k P_l P_{k'} P_{l'}}\, dt$$
$$= \sum_{k,l} k^2 P_k P_l\, dt - \sum_{k,l,k',l'} kl P_k P_l\, dt$$
$$= (m_2 - m^2)\, dt$$
$$= c_2\, dt.$$

Shiga (1982) showed that although there does not exist a stationary distribution for the neutral process $P(t)$, there does exist a stationary distribution for the process

1 Rate of Adaptation of Large Populations

centred about its mean, and furthermore, the centred process is ergodic, i.e. after a long time, the process behaves like the stationary distribution. Therefore, we expect the expectation of all the central moments of P to converge to something as $t \to \infty$. To find the exact value of $\lim_{t \to 0} E^{\mathbb{P}_0}[c_n(P(t))]$, $n = 2, 3, \ldots$, we use the generator $\mathcal{A}_{0,q,\tilde{\mu}}$ to calculate its effect on c_n. First, it can be verified that for $n \geq 2$,

$$\frac{\partial c_n}{\partial p_k} = (k-m)^n - knc_{n-1}$$

$$\frac{\partial^2 c_n}{\partial p_k \partial p_l} = -ln(k-m)^{n-1} - nk((l-m)^{n-1} - (n-1)lc_{n-2}),$$

where $c_1 = 0$ and $c_0 = 1$ and $m = m(p)$. Plugging these two formulae into the definition of $\mathcal{A}_{0,q,\tilde{\mu}}$ in Eq. 1.9, we obtain

$$\mathcal{A}_{0,q,\tilde{\mu}} c_n = \tilde{\mu} \sum_k [q(k-m+1)^n - (k-m)^n - (1-q)(k-m-1)^n] p_k$$

$$- \tilde{\mu}(2q-1)nc_{n-1} + \frac{1}{2}n(n-1)c_2 c_{n-2} - nc_n$$

$$= \frac{1}{2}n(n-1)c_2 c_{n-2} - nc_n + \tilde{\mu} \sum_{i=2}^n \binom{k}{i}(q + (-1)^i(1-q))c_{n-i}.$$

In particular,

$$\mathcal{A}_{0,q,\tilde{\mu}} c_2 = \tilde{\mu} - c_2$$
$$\mathcal{A}_{0,q,\tilde{\mu}} c_3 = \tilde{\mu}(2q-1) - 3c_3$$
$$\mathcal{A}_{0,q,\tilde{\mu}} c_4 = \tilde{\mu}(6c_2 + 1) + 6c_2^2 - 4c_4$$
$$\mathcal{A}_{0,q,\tilde{\mu}} c_5 = \tilde{\mu}(10c_3 + 10(2q-1)c_2 + (2q-1)) + 10c_3 c_2 - 5c_5, \quad (1.21)$$

and Itô's formula yields

$$c_n(P(t)) = c_n(P(0)) + \int_0^t \mathcal{A}_{0,q,\tilde{\mu}} c_n(P(s)) \, ds$$

$$+ \int_0^t \sum_{k,l} [(k - m(P(s)))^n - knc_{n-1}(P(s))] \sqrt{P_k(s) P_l(s)} \, dW_{kl}(s), \quad (1.22)$$

where the last integral is a martingale. For c_2, we take expectation on both sides of Eq. 1.22 to obtain

$$E^{\mathbb{P}_0}[c_2(P(t))] = E^{\mathbb{P}_0}[c_2(P(0))] + \int_0^t (\tilde{\mu} - E^{\mathbb{P}_0}[c_2(P(s))]) \, ds.$$

In differential form, we can write

$$\frac{d}{dt} E^{\mathbb{P}_0}[c_2(P(t))] = \tilde{\mu} - E^{\mathbb{P}_0}[c_2(P(t))].$$

Thus, $E^{\mathbb{P}_0}[c_2(P(t))] = E^{\mathbb{P}_0}[c_2(P(0))]e^{-t} + \tilde{\mu}(1-e^{-t})$, and hence

$$\lim_{t\to\infty} E^{\mathbb{P}_0}[c_2(P(t))] = \tilde{\mu}. \tag{1.23}$$

Similarly,

$$\lim_{t\to\infty} E^{\mathbb{P}_0}[c_3(P(t))] = \frac{\tilde{\mu}}{3}(2q-1). \tag{1.24}$$

For $n=4$ and $n=5$, $\mathcal{A}_{0,q,\tilde{\mu}}c_n$ involves terms of the form $c_{n-2}c_2$ in addition to c_n:

$$\mathcal{A}_{0,q,\tilde{\mu}}(c_2^2) = 2\tilde{\mu}c_2 + c_4 - 3c_2^2,$$
$$\mathcal{A}_{0,q,\tilde{\mu}}(c_3c_2) = \tilde{\mu}[c_3 + (2q-1)c_2] + c_5 - 6c_3c_2.$$

Combining the above two formulae with Eq. 1.21 yields

$$\lim_{t\to\infty} E^{\mathbb{P}_0}[c_4(P(t))] = 5\tilde{\mu}\lim_{t\to\infty} E^{\mathbb{P}_0}[c_2(P(t))] + \frac{\tilde{\mu}}{2} = 5\tilde{\mu}^2 + \frac{\tilde{\mu}}{2},$$

$$\lim_{t\to\infty} E^{\mathbb{P}_0}[c_2^2(P(t))] = \frac{7}{3}\lim_{t\to\infty} E^{\mathbb{P}_0}[c_2(P(t))] + \frac{\tilde{\mu}}{2} = \frac{7}{3}\tilde{\mu}^2 + \frac{\tilde{\mu}}{2},$$

$$\lim_{t\to\infty} E^{\mathbb{P}_0}[c_5(P(t))] = \tilde{\mu}(2q-1)\left(\frac{5\tilde{\mu}}{3} + \frac{3}{20}\right),$$

$$\lim_{t\to\infty} E^{\mathbb{P}_0}[c_3(P(t))c_2(P(t))] = \tilde{\mu}(2q-1)\left(\frac{\tilde{\mu}}{2} + \frac{1}{40}\right). \tag{1.25}$$

Lastly, we need the following result

$$d[M, c_n] = \sum_{k,l,k',l'} ((k-m)^n - nkc_{n-1})k'\sqrt{P_k P_l P_{k'} P_{l'}}\, d[W_{kl}, W_{k'l'}]$$

$$= \left(\sum_k k((k-m)^n - nkc_{n-1})P_k - \sum_{k,l}((k-m)^n - nkc_{n-1})lP_kP_l\right) dt$$

$$= \sum_k (k-m)((k-m)^n - nkc_{n-1})P_k\, dt$$

$$= (c_{n+1} - nc_{n-1}c_2)\, dt. \tag{1.26}$$

1.4.3 Representation Using the Neutral Process

From Eq. 1.8, the mean of the selected process $\tilde{P}$ satisfies

$$m(\tilde{P}(t)) = m(\tilde{P}(0)) + \int_0^t \sum_k k[\tilde{\mu}(q\tilde{P}_{k-1}(s) - \tilde{P}_k(s) + (1-q)\tilde{P}_{k+1}(s))$$
$$+ \tilde{\sigma}(k - m(\tilde{P}(s)))\tilde{P}_k(s)]\, ds + M(t)$$
$$= \tilde{\mu}(2q-1)t + \tilde{\sigma}\int_0^t c_2(\tilde{P}(s))\, ds + M(t). \tag{1.27}$$

This is similar to Eq. 1.4, i.e. the strong selection case of model A, where the rate of adaptation is also proportional to the variance of fitness. The critical quantity of interest is the integrated variance $\int_0^t c_2(\tilde{P}(s))\, ds$. But if we try to write down the evolution equation of c_2, we will again obtain a system resembling Eq. 1.12.

For the weak selection model, however, we have a representation of the selected process $\mathcal{A}_{\tilde{\sigma},q,\tilde{\mu}}$ in terms of the neutral process, and we can write central moments of the selected process using this representation. For example, let $P(t)$ be the solution to the neutral SDE (Eq. 1.15), then the integrated variance of the selected process associated with $\mathcal{A}_{\tilde{\sigma},q,\tilde{\mu}}$ can be written as

$$E^{\mathbb{P}_{\tilde{\sigma}}}\left[\int_0^t c_2(\tilde{P}(s))\, ds\right] = E^{\mathbb{P}_0}\left[Z(t)\int_0^t c_2(P(s))\, ds\right]. \quad (1.28)$$

We define

$$N(t) = \int_0^t c_2(P(s))\, ds$$

and write

$$Z(t) = \exp\left(\tilde{\sigma} M(t) - \frac{\tilde{\sigma}^2}{2} N(t)\right).$$

We expand Z into a Taylor series around 0 and write the term inside the expectation on the right-hand side of Eq. 1.28 as follows, dropping the dependence on t of M and N in our notation:

$$Z(t)N = \sum_{k=0}^{\infty} \frac{1}{k!}\left(\tilde{\sigma} M - \frac{\tilde{\sigma}^2}{2} N\right)^k N.$$

We can write the left-hand side of Eq. 1.28 as an expansion in the selection coefficient $\tilde{\sigma}$:

$$\begin{aligned}
E^{\mathbb{P}_{\tilde{\sigma}}}\left[\int_0^t c_2(\tilde{P}(s))\, ds\right] &= E^{\mathbb{P}_0}\left[\sum_{k=0}^{\infty} \frac{1}{k!}\left(\tilde{\sigma} M - \frac{\tilde{\sigma}^2}{2} N\right)^k N\right] \\
&= E^{\mathbb{P}_0}\left[N + \left(\tilde{\sigma} M - \frac{\tilde{\sigma}^2}{2} N\right) N + \frac{1}{2}\left(\tilde{\sigma} M - \frac{\tilde{\sigma}^2}{2} N\right)^2 N\right. \\
&\quad \left. + \frac{1}{6}\left(\tilde{\sigma} M - \frac{\tilde{\sigma}^2}{2} N\right)^3 N + \ldots\right] \\
&= E^{\mathbb{P}_0}[N] + \tilde{\sigma} E^{\mathbb{P}_0}[MN] + \frac{\tilde{\sigma}^2}{2} E^{\mathbb{P}_0}\left[(M^2 - N)N\right] \\
&\quad + \frac{\tilde{\sigma}^3}{6} E^{\mathbb{P}_0}\left[(M^3 - 3MN)N\right] + \ldots,
\end{aligned}$$

where in the last step we have taken the expectation inside the infinite sum, which needs to be justified rigorously. Assuming we can do this, we can quite easily compute the approximate value of the first few terms in the above expansion by hand for large time and they yield some interesting information about the rate of adaptation of the selected process $\mathcal{A}_{\tilde{\sigma},q,\tilde{\mu}}$. For the first term in the expansion

$$E^{\mathbb{P}_0}[N] = E^{\mathbb{P}_0}\left[\int_0^t c_2(P(s))\,ds\right] = \int_0^t E^{\mathbb{P}_0}[c_2(P(s))]\,ds,$$

since $\lim_{t\to\infty} E^{\mathbb{P}_0}[c_2(P(t))] = \tilde{\mu}$ by (1.23), N increases at rate $\tilde{\mu}$ for large t.

Now we compute the second term in the expansion $E^{\mathbb{P}_0}[MN]$.

$$d(MN) = M\,dN + d\text{martingale} = Mc_2\,ds + d\text{martingale}.$$

Thus, as $t \to \infty$, $E^{\mathbb{P}_0}[M(t)N(t)]$ increases at a rate of $\lim_{t\to\infty} E^{\mathbb{P}_0}[M(t)c_2(t)]$.

$$\begin{aligned}d(Mc_2) &= M\,dc_2 + d[M,c_2] + d\text{martingale} \\ &= (-Mc_2 + c_3)\,dt + d\text{martingale}\end{aligned}$$

by Eq. 1.26. Taking expectation on both sides, we get

$$E^{\mathbb{P}_0}[M(t)c_2(P(t))] = \int_0^t E^{\mathbb{P}_0}[c_3(P(s))] - E^{\mathbb{P}_0}[M(s)c_2(P(s))]\,ds.$$

By an argument similar to that leading to Eq. 1.23, we should expect

$$\lim_{t\to\infty} E^{\mathbb{P}_0}[M(t)c_2(P(t))] = \lim_{t\to\infty} E^{\mathbb{P}_0}[c_3(P(t))] = \frac{\tilde{\mu}}{3}(2q-1),$$

by Eq. 1.24. Thus, $E^{\mathbb{P}_0}[MN]$ increases at a rate of $\frac{\tilde{\mu}}{3}(2q-1)$ as $t \to \infty$.

For the third term in the expansion $E^{\mathbb{P}_0}[(M^2 - N)N]$, we first observe that $M(t)^2 - N(t)$ is a martingale and can be written as $\int_0^t 2M(s)\,dM(s)$. Therefore

$$\begin{aligned}d((M^2-N)c_2) &= (M^2-N)\,dc_2 + d[M^2-N,c_2] + d\text{martingale} \\ &= -(M^2-N)c_2\,dt + 2M\,d[M,c_2] + d\text{martingale} \\ &= [-(M^2-N)c_2 + 2Mc_3]\,dt + d\text{martingale}.\end{aligned}$$

by Eq. 1.26. So we need to calculate $\lim_{t\to\infty} E^{\mathbb{P}_0}[M(t)c_3(P(t))]$:

$$\begin{aligned}d(Mc_3) &= M\,dc_3 + d[M,c_3] + d\text{martingale} \\ &= (-3Mc_3 + c_4 - 3c_2^2)\,dt + d\text{martingale},\end{aligned}$$

by Eqs. 1.21 and 1.26. From Eq. 1.25, we have

$$\lim_{t\to\infty} E^{\mathbb{P}_0}[c_4(P(t)) - 3c_2(P(t))^2] = -2\tilde{\mu}^2 - \tilde{\mu},$$

which means that

$$\lim_{t\to\infty} E^{\mathbb{P}_0}[(M(t)^2 - N(t))c_2(P(t))] = 2\lim_{t\to\infty} E^{\mathbb{P}_0}[M(t)c_3(P(t))]$$
$$= \frac{2}{3}\lim_{t\to\infty} E^{\mathbb{P}_0}[c_4(P(t)) - 3c_2(P(t))^2]$$
$$= -\frac{2}{3}(2\tilde{\mu}^2 + \tilde{\mu}).$$

So we should expect $E^{\mathbb{P}_0}\left[(M^2 - N)N\right]$ to increase at a rate of $-\frac{2}{3}(2\tilde{\mu}^2 + \tilde{\mu})$ as $t \to \infty$.

Similarly, we can calculate that $E^{\mathbb{P}_0}\left[(M^3 - 3MN)N\right]$ increases at a rate of $4\tilde{\mu}(2q-1)(\frac{\tilde{\mu}}{6} + \frac{1}{40})$ as $t \to \infty$. Thus, we can write down the first few terms of the expansion of the left-hand side of Eq. 1.28 for large t

$$\tilde{\mu}t + \tilde{\sigma}\frac{\tilde{\mu}}{3}(2q-1)t - \tilde{\sigma}^2\frac{1}{3}(2\tilde{\mu}^2 + \tilde{\mu})t + \tilde{\sigma}^3\frac{2}{3}(2q-1)\left(\frac{\tilde{\mu}^2}{6} + \frac{\tilde{\mu}}{40}\right)t + \cdots.$$

Plugging this result into Eq. 1.27, we see that we can write the limiting adaptation rate of the process (Eq. 1.8) as an expansion in $\tilde{\sigma}$, where the first few terms are as follows:

$$\tilde{\mu}(2q-1) + \tilde{\sigma}\tilde{\mu} + \tilde{\sigma}^2(2q-1)\frac{\tilde{\mu}}{3} - \tilde{\sigma}^3\left(\frac{2}{3}\tilde{\mu}^2 + \frac{\tilde{\mu}}{3}\right)$$
$$+ \tilde{\sigma}^4(2q-1)\left(\frac{\tilde{\mu}^2}{9} + \frac{\tilde{\mu}}{60}\right) + \cdots. \quad (1.29)$$

The first term above is simply the adaptation rate for the neutral process $\mathcal{A}_{0,q,\tilde{\mu}}$. The second term can be interpreted as the first-order correction on the adaptation rate assuming the selected process $\mathcal{A}_{\tilde{\sigma},q,\tilde{\mu}}$ has the same variance of fitness $\tilde{\mu}$ as the neutral process $\mathcal{A}_{0,q,\tilde{\mu}}$. We expect the selected process to have a smaller variance of fitness than the neutral process. But the second-order correction (the $\tilde{\sigma}^2$ term) actually increases the variance if $2q - 1 > 0$. If $2q - 1 < 0$, on the other hand, then all correction terms on the variance of fitness are negative, as expected. It is the third-order correction (the $\tilde{\sigma}^3$ term) that starts to reduce the variance of fitness in the case of $2q - 1 > 0$.

1.5 Open Problems

Although we have laid out in Sect. 1.4.3 a way to write the adaptation rate in the form of a series in the selection coefficient $\tilde{\sigma}$, and demonstrated how, in principle, one can calculate every term in this series, any term above $\tilde{\sigma}^5$ becomes very tedious to calculate by hand. For the series to be of use for realistic ranges of $\tilde{\sigma}$, say 10–100, an efficient way of calculating higher-order terms in the series (Eq. 1.29) becomes necessary. The selection coefficient $\tilde{\sigma}$ in the weak selection model roughly corresponds with σN in the strong selection model. As N becomes larger, the strong selection model becomes a better approximation. But how large does N need to get for the strong selection model to be a reasonable approximation? From simulations, $\sigma N \geq 500$ seems to be sufficient. Is there a reasonable way to quantitatively measure how good or bad the strong selection model is, so that we obtain some sort of confidence interval for the prediction on adaptation rate we obtain using this model?

Acknowledgements The authors are grateful to Nick Barton for posing the question of whether there exists an upper limit to the adaptation rate in the asexual case. We are also grateful to Nick Barton and Jonathan Coe for valuable discussions throughout this research. In addition, we thank Charles Cuthbertson as a co-author of Cuthbertson et al. (2007). F.Y. was supported by EPSRC/GR/T19537 while at the University of Oxford.

References

Barton NH, Coe JB (2007) An upper limit to the rate of adaptation. Preprint
Cuthbertson C, Etheridge AM, Yu F (2007) Asymptotic behaviour of the rate of adaptation. arXiv:0708.3453
Dawson DA (1993) Measure-valued Markov processes. In: École d'Été de Probabilités de Saint-Flour XXI—1991. Lecture notes in mathematics, vol 1541. Springer, Berlin, pp 1–260
Desai MM, Fisher DS (2007) Beneficial mutation selection balance and the effect of linkage on positive selection. Genetics 176:1759–1798
Donnelly P, Kurtz TG (1999) Genealogical processes for Fleming–Viot models with selection and recombination. Ann Appl Probab 9(4):1091–1148
Etheridge AM, Pfaffelhuber P, Wakolbinger A (2007) How often does the ratchet click? Facts, heuristics, asymptotics. Trends Stochastic Anal. arXiv:0709.2775v1
Etheridge AM (2000) An introduction to superprocesses. University lecture series, vol 20. American Mathematical Society, Providence
Fisher RA (1930) The genetical theory of natural selection. Clarendon, Oxford
Gerrish PJ, Lenski RE (1998) The fate of competing beneficial mutations in an asexual population. Genetica 102/103:127–144
Gillespie JH (1991) The causes of molecular evolution. Oxford University Press, Oxford
Gordo I, Charlesworth B (2000) On the speed of Muller's ratchet. Genetics 156:2137–2140
Haigh J (1978) The accumulation of deleterious genes in a population. Theor Popul Biol 14(2):251–267
Haldane JBS (1927) A mathematical theory of natural and artificial selection, part V: selection and mutation. Proc Camb Philos Soc 23:834–844
Hegreness M, Shoresh N, Hartl D, Kishony R (2006) An equivalence principle for the incorporation of favourable mutations in asexual populations. Science 311:1615–1617
Higgs P, Woodcock G (1995) The accumulation of mutations in asexual populations, and the structure of genealogical trees in the presence of selection. J Math Biol 33:677–702
Ikeda N, Watanabe S (1981) Stochastic differential equations and diffusion processes. North-Holland mathematics library. North-Holland, Amsterdam
Karatzas I, Shreve SE (1991) Brownian motion and stochastic calculus, 2nd edn. Graduate texts in mathematics, vol 113. Springer, New York
Novella IS, Elena SF, Moya A, Domingo E, Holland JJ (1995) Size of genetic bottlenecks leading to virus fitness loss is determined by mean initial population fitness. J Virol 69:2869–2872

Novella IS, Elena SF, Moya A, Domingo E, Holland JJ (1999) Exponential fitness gains of RNA virus populations are limited by bottleneck effects. J Virol 73:1668–1671

Rouzine I, Brunet E, Wilke CO (2007) The traveling wave approach to asexual evolution: Muller's ratchet and speed of adaptation. arXiv:0707.3469

Rouzine I, Wakeley J, Coffin JM (2003) The solitary wave of asexual evolution. Proc Natl Acad Sci USA 100(2):587–592

Shiga T (1982) Wandering phenomena in infinite-allelic diffusion models. Adv Appl Probab 14(3):457–483

Stephan W, Chao L, Smale J (1993) The advance of Muller's ratchet in a haploid asexual population: approximate solution based on diffusion theory. Genet Res 61:225–232

Wilke CO (2004) The speed of adaptation in large asexual populations. Genetics 167:2045–2054

Chapter 2
A Phylogenetic Mixture Model for Heterotachy

Andrew Meade and Mark Pagel

Abstract We present a likelihood-based phylogenetic mixture model designed to analyse data that exhibit within-site rate variation or heterotachy. Heterotachy refers to the phenomenon of a site in a gene-sequence or other alignment changing its rate of evolution throughout the tree. The method accounts for heterotachy by summing the likelihood of the data at each site over more than one set of branch lengths on the same tree. A branch length set that is best for one site may differ from the branch length set that is best for some other site, thereby allowing different sites to have different rates of change throughout the tree. We show that the model improves the accuracy of phylogenetic reconstruction when the sequence data are not derived from a single underlying evolutionary process. We apply the method to a number of simulated and published data sets and show that many sequence data sets have complex evolutionary signals of heterotachy. The presence of such signals has important consequences for the correct reconstruction of phylogenies as well as for tests of hypotheses that rely on accurate branch length information. These include molecular clocks, analyses of tempo and mode of evolution, comparative studies and ancestral state reconstruction. The model is implemented in a Bayesian Markov chain Monte Carlo framework and is available from the authors' Web site, and can be used for the analysis of both nucleotide and morphological data.

2.1 Introduction

Multigene phylogenies have become the standard in systematic biology, and different molecular sequences often have their own mode and tempo of evolution. Some are highly conserved on a geological time frame, others undergo rapid mutation,

---

A. Meade and M. Pagel
School of Biological Sciences, Philip Lyle Building, The University of Reading,
Reading RG6 6BX, UK
A.Meade@reading.ac.uk, m.pagel@reading.ac.uk

evolving on a human cultural scale. These properties have been used to great effect by researchers reconstructing phylogentic trees. By carefully choosing combinations of sequences, phylogenies can accurately trace the evolution of both closely related species as well as deeper nodes representing distant relationships. Where resolution is poor, additional sequences are employed to untangle complex evolutionary processes.

Combining sequences from different evolutionary processes can also introduce error into phylogentic reconstruction methods, especially methods that assume a single homogeneous process operating throughout the tree. Even when single-gene sequence information is used to reconstruct phylogentic trees, the topology and branch lengths represent some average solution to characterising the evolution of that gene. When multiple sequences are combined, the topology and branch lengths are forced to represent an average over genes that may differ greatly in their patterns and rates of change. Combining gene sequences from different evolutionary processes can lead to the incorrect topology being reconstructed even when all the sequences have evolved on the same topology (Kolaczkowski and Thornton 2004). This process can even be positively misleading, such that with the addition of more data the incorrect outcome appears more certain.

'Heterotachy' refers to the phenomenon of a site's rate of evolution varying throughout its evolutionary history. It is probably prevalent and widespread throughout genetic loci. For example, one study of mitochondrial cytochrome b, found that 95% of variable positions exhibited some degree of heterotachy (Lopez et al. 2002), and it has been identified in mammalian promoter regions (Taylor et al. 2006) and in plastid genes (Shalchian-Tabrizi et al. 2006). The causes of heterotachy are undoubtedly complex, but changes to selection pressures, the environment, functional constraints and the knock-on effect of gene transfer have been suggested (Ane et al. 2005; Lopez et al. 2002; Taylor et al. 2006). Heterotachy is known to affect phylogenetic reconstruction (Huelsenbeck 2002; Kolaczkowski and Thornton 2004; Lopez et al. 2002), being a source of error to both maximum parsimony and likelihood based models (Philippe et al. 2005). One of its more pronounced effects is long-branch attraction (Philippe et al. 2005).

Models governing protein and DNA sequence evolution are fundamental to molecular analysis (Yang 1994). The accuracy of phylogenetic inferences is greatly improved when models accounting for variations in evolutionary rates are employed. Two common and elegant methods are the gamma rate heterogeneity model (Yang, 1994) and covarion models (Fitch and Markowitz 1970; Galtier 2001; Huelsenbeck 2002; Tuffley and Steel 1998). Gamma rate heterogeneity integrates the rate of change, on a site-by-site basis, over a discretised gamma distribution, fitting slowly evolving sites with a low rate of change and faster-evolving sites with a higher rate. Gamma rate heterogeneity has become the de facto model for across-site rate variation, enabling the rate of evolution to change across sites but to remain constant with in a site.

The covarion model (e.g. Tuffley and Steel) allows the rate of change to vary within a site throughout the tree. Covarion models switch between an 'on' and an 'off' state implemented in a hidden Markov process. This allows subsections of the tree or even individual branches to be turned on and off for a given site as it evolves in the tree. An effective modification that allows a number of hidden states with varying rates, instead of 'on' and 'off', was proposed by Galtier (2001). While elegant, covarion models assume a parametric structure of switches between the on and the off state that is assumed to apply to all sites. If this parametric switching structure is not true of the sites, covarion models may only return modest improvements over homogeneous models applied to the same heterotachous data (Kolaczkowski and Thornton 2004).

Mixture models (Gelman 2003) are a class of methods for fitting more than one statistical model to a set of data. In a phylogentic context, a mixture model might sum the likelihood of the data over a range of models of sequence evolution. Mixture models are useful for describing data that are drawn from more then one process, and they have been successfully developed for phylogenetic inference (Huelsenbeck 1999; Koshi and Goldstein 1998; Pagel and Meade 2004). In previous work we have developed mixture models for characterising data in which different sites have evolved according to different models of sequence evolution (Pagel and Meade 2004, 2005). We call this 'pattern heterogeneity' and we have shown that it often outperforms partitioned analyses of the same data. Pattern-heterogeneity models have proved to be effective in reducing node density effects and long branch attraction (Philippe et al. 2005; Venditti et al. 2008), and frequently yield longer trees, suggesting that they account for more evolution.

Here we present a mixture model to account for heterotachy within sequence data. The model sums the likelihood of the data over a number of independent branch length sets on the same tree. If two branch length sets differ in the length of a branch at a given place in the tree, this corresponds to faster and slower rates of evolution along that branch. The use of independent branch length sets therefore allows different sites to have different rates of evolution in different parts of the tree, but without assuming any parametric structure to changes in the rates of evolution. We test the model in four-taxon tree for a scenario in which parsimony has been shown to outperform conventional likelihood methods (Kolaczkowski and Thornton 2004). A second simulation study is used to show that the correct number of branch length sets can be recovered without prior knowledge of the number present in the data. We then report application of the branch lengths model to a number of published data sets.

2.2 Branch Length Sets Mixture Model

Define the likelihood of a model of gene sequence evolution as an amount proportional to the probability of the data given the model:

$$L(\mathbf{Q}) \propto\leftarrow P(\mathbf{D}|\mathbf{Q}), \tag{2.1}$$

where $\mathbf{Q}$ is the substitution rate matrix that defines the model of evolution, and $\mathbf{D}$ will normally be an aligned set of sequence data. In the case of nucleotide data, $\mathbf{Q}$ is the familiar 4 × 4 matrix of transition rates among A, C, G and T (Swofford et al. 1996). For protein data **Q** is a 20 × 20 matrix representing the transition rates among all pairs of amino acids.

Given an aligned set of gene-sequence or other character-state data, the probability of the data in **D** is found as the product over sites of the individual probabilities of each site. Considering that the likelihood is calculated for a specific phylogenetic tree, we can write the right-hand side of Eq. 2.1 as

$$P(\mathbf{D}|\mathbf{Q},T) = \prod_i P(\mathbf{D}_i|\mathbf{Q},T), \qquad (2.2)$$

where the product is over all of the sites in the data matrix and T stands for the specific tree.

A mixture model for additional branch length sets modifies this basic framework by including more than one set of branches for a given tree T. The branching structures (topology) of the trees are fixed, while independent branch lengths represented as a vector **t** are free to vary. The probability of the data is now calculated by summing the likelihood at each site over all of the different **t** for a given tree. Thus, defining the branch length sets as $\mathbf{t}_1, \mathbf{t}_2, \ldots, \mathbf{t}_J$, write the probability of the data under the mixture model as

$$P(\mathbf{D}|\mathbf{t}_1,\mathbf{t}_2,\ldots,\mathbf{t}_J,\mathbf{Q},T) = \prod_i \sum_j w_j P(\mathbf{D}_i|\mathbf{t}_j,\mathbf{Q},T), \qquad (2.3)$$

where the summation over j now specifies that the likelihood of the data at each site is summed over J separate branch-length sets T, the summation being weighted by w's, where $w_1, w_2, \ldots, w_J = 1.0$. The number of branch length sets, J, can be determined either by prior knowledge of how many different patterns are expected in the data, or it can be empirically estimated from the data.

An earlier study (Kolaczkowski and Thornton 2004) presented a mixture model with two branch length sets but calculated the likelihood in Eq. 2.3 incorrectly (Spencer et al. 2005) and it was only applicable to data in which it was known a priori which branch length set best characterised a given site.

2.3 Model Testing

We employ the Akaike information criterion (AIC) to select between models (Akaike 1974; Felsenstein 2004) with different numbers of parameters. Models with a larger number of parameters are expected to achieve a higher likelihood. The AIC value is calculated for each model, and the model with the lowest AIC value is selected:

$$\mathrm{AIC} = -2\,\mathrm{Ln}_i + 2p_i,$$

where Ln_i is the log-likelihood for model i, and p_i is the number of parameters. The addition of each branch length set requires $2n$-3 extra branches for an unrooted tree,

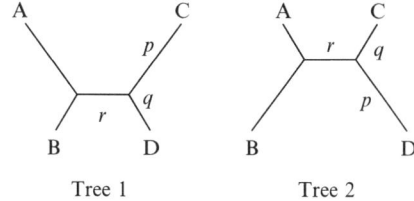

Fig. 2.1 Two topologies used to simulate sequence evolution

where n is the number of taxa. In addition the model estimates a weighting term for each set. As the models are implemented in a Bayesian framework we use the mean of the posterior distribution of log-likelihood for Ln.

2.4 Four-Taxon Simulation

Following Kolaczkowski and Thornton (2004), we simulated data for a four-taxon tree in which individual sites evolve according to one of two separate branch length sets.

We used Seq-Gen (Rambaut and Grassly 1997) to simulate 5,000 random data sets using the two sets of branches shown in Fig. 2.1. We varied r between 0 and 0.4 in 5,000 steps, with p and q constant at, 0.75 and 0.05, respectively. Five hundred nucleotides were simulated for each set of branches using the Jukes–Cantor model with equal base frequencies, and were combined into a single data set of 1,000 sites. Because of the wide variation in branch lengths between the two halves of the data, we expect homogeneous models to suffer from long-branch attraction. To account for heterotachy, the data sets were analysed using the model in Eq. 2.3, implemented in a Markov chain Monte Carlo (MCMC) framework (Geyer 1992) The method is available in the software package in BayesPhylogenies (Pagel and Meade 2004). Each data set was run estimating a single set of branches to mimic a homogeneous model and a mixture model that estimated two sets of branches. We sampled 500 trees from the Markov chain after appropriate convergence, at intervals of 2,500 iterations to ensure independence among successive samples.

2.5 Seventy-Taxon Simulation

We conducted an additional study to determine whether the correct number of branch length sets and their weights could be recovered from simulated data. We used a random topology of 70 taxa, and created four independent sets of branches for this topology. Each set of branches was derived by drawing random lengths from a uniform distribution ranging between 0 and 0.2. We then generated 1,500 simulated nucleotides for each branch length set, using Seq-Gen (Rambaut and Grassly 1997). In every case we used the same general time-reversible (GTR) model of evolution.

We then combined two of the branch length sets to make a single alignment of 3,000 simulated nucleotides. A second alignment was derived from three of the simulated data sets and a third was based on all four (Table 2.1).

We analysed each simulated alignment using a conventional GTR likelihood model with one set of branches, and also with the branch length set (BLS) mixture model using two, three and four distinct branch length sets. We ran four independent non-mixing chains for each analysis to identify the stationary distribution. We found that 1,000,000 iterations gave the chain ample time for apparent convergence and then we sampled every 10,000 iterations such that successive samples were reasonably independent.

2.6 Application to Published Data Sets

We applied the model to the five published data sets described in Table 2.2. Data sets were taken from TreeBase (Piel et al. 2002), with the corresponding reference shown in parentheses. Each of these alignments contains more than one gene. We fit one, two and three branch length sets to these data sets (in no cases was a fourth needed). A period of 1,000,000 iterations was used as burn-in and the chain was sampled every 10,000 iterations. Yang's (1994) gamma rate heterogeneity model was applied to all data sets with the gamma rate parameter estimated across branch length sets (thus all results are independent of rate heterogeneity across sites). Three independent chains were used to check for the stationary distribution.

Table 2.1 The grouping of four simulated branch length sets into three alignments

Simulated branch length sets[a]	No. of sites in simulated alignment
2	3,000
3	4,500
4	6,000

[a] See the text for a description of the method.

Table 2.2 Studies, taxa number and aligned sites for five published data sets

Taxa	Genes	No. of taxa	No. of sites
Chlorophyceae (Buchheim et al. 2001)	18*S* and 26*S*	38	3,684
Gnetales (Rydin et al. 2002)	26*s*, 18*s*, *rbcL* and *atpB*	119	5,923
Caenorhabditis phylogeny (M1869) (Kiontke et al. 2004)	18*s*, 28*s*, *RNAP2*, *Par6* and *pkc3*	14	7,652
Plethodontid salamanders (Mueller et al. 2004)	Complete mitochondrial genomes	27	14,040
Costaceae (Zingiberales) (Specht 2006)	*ITS*, *trml-F*, *trnK* and *matK*	66	5,898

2.7 Results

2.7.1 Four-Taxon Simulation

The four-taxon simulation results are shown in Fig. 2.2, where we plot the probability of recovering the true topology shown in Fig. 2.1 against the length of the internal branch r. We characterise the results using a non-linear logistic growth curve. The model using one branch length set required an r of 0.285 to achieve a 0.5 probability of recovering the true tree. This corresponds to a conventional likelihood model. The model with two branch length sets required an r of just 0.04 to meet the same criterion. Maximum parsimony requires an r of 0.22 (Kolaczkowski and Thornton 2004) to recover the true tree with a probability 0.5. Even though the true tree is used to simulate all sites, a simple model, unable to account for heterotachy, fails to determine the correct tree when the internal branch is short. It fails by placing A with C and B with D until r becomes sufficiently long to separate them. When r is less then 0.15 the posterior probability for the incorrect result for the homogeneous model is 1.0. Both rate heterogeneity and covarion models (not shown) did not improve upon the one branch length set results.

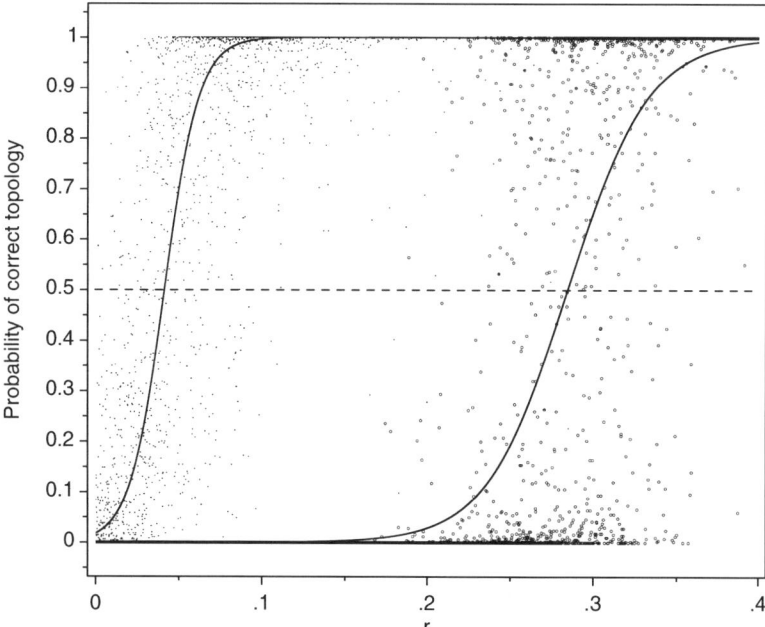

Fig. 2.2 Probability of recovering the correct topology plotted on the y-axis, with the internal branch length r on the x-axis, see Fig. 2.1. *Circles* plot results using a single branch length set, *squares* use two branch length sets. The logistic $2p$ model was used as the line of best fit for each model. The r value needed for 0.5 probability of recovering the correct topology is 0.285 for one branch length set and 0.04 for two. Results from applying gamma and covarion models are identical to those for one branch length set

Table 2.3 Results of the 70-taxon simulation study. The rows in *bold* represents the model used to simulate the data and the model of choice for analysis testing using the Akaike information criterion (AIC)

Simulated branch length sets	Branch length set model	log-L ± SD	Weight 1	Weight 2	Weight 3	Weight 4
2	1	−131,951 ± 8.937	1			
2	**2**	**−131,028 ± 12.026**	**0.513**	**0.487**		
2	3	−131,029 ± 13.094	0.509	0.49	0	
2	4	−131,028 ± 15.887	0.508	0.485	0.007	0
3	1	−198,452 ± 8.242	1			
3	2	−197,783 ± 11.606	0.631	0.369		
3	**3**	**−196,898 ± 14.498**	**0.349**	**0.33**	**0.321**	
3	4	−196,907 ± 18.75	0.353	0.332	0.315	0
4	1	−266,093 ± 8.714	1			
4	2	−265,210 ± 12.403	0.759	0.241		
4	3	−264,513 ± 16.075	0.502	0.254	0.244	
4	**4**	**−263,793 ± 17.895**	**0.277**	**0.25**	**0.237**	**0.235**

log-L mean log-likelihood, *SD* standard deviation.

2.7.2 Seventy-Taxon Simulation

Table 2.3 shows the results of the 70-taxon simulation. The branch lengths model can find the correct number of branch lengths sets in a simulated alignment and it is able to estimate the weights to assign to each branch length set. The weights represent the proportion of sites associated with each branch length set. The table also shows that the correct model is favoured in each case by the AIC test. Thus, when applied to the simulated alignment with two branch length sets, the second branch length set improved the likelihood by more than 900 log units compared with a single set of branches, and the estimated weight for each branch length set ($w = 0.5$) was recovered accurately (see Table 2.3, weight columns). The addition of a third and fourth branch length set to these data did not lead to a significant likelihood improvement and the estimated weight associated with the unneeded branch length sets was very low, indicating that the additional branches have little or no effect. This shows that if the model is overparameterised (i.e. too many branch length sets) there will be insufficient likelihood improvement and a low weight associated with the unsupported extra branches. The other two alignments show the same pattern.

2.7.3 Application to Real Data

Table 2.4 shows the results from applying the model to the five real data sets. Under the AIC each data set supported more than one set of branches and the data from Mueller et al. (2004) support three sets.

Table 2.4 The results of applying models with one, two and three branch length sets to five published data sets. Each data set was tested using the AIC and the selected model for each is highlighted in *bold*

Data	Branch length sets in model	log-L ± SD	Parameters[a]	log-L gain	AIC	w_1	w_2	w_3
Chlorophyceae	1	−26,260.1 ± 20.18	0	0	52,520.2	1	–	–
	2	**−26,147.9 ± 9.57**	**74**	**112.2**	**52,443.7**	**0.56**	**0.44**	**–**
	3	−26,138.6 ± 12.07	148	121.5	52,573.2	0.82	0.11	0.07
Gnetales	1	−77,650 ± 13.9	0	0	155,300.1	1	–	–
	2	**−77,145.2 ± 44.64**	**236**	**504.8**	**154,762.3**	**0.39**	**0.61**	**–**
	3	−77,116.3 ± 14.15	472	533.7	155,176.6	0.56	0.39	0.06
Caenorhabditis	1	−43,639.9 ± 5.99	0	0	87,279.9	1	–	–
	2	**−43,516.3 ± 6.58**	**26**	**123.6**	**87,084.6**	**0.61**	**0.39**	**–**
	3	−43,502.9 ± 7.65	52	137	87,109.8	0.56	0.3	0.14
Plethodontid salamanders	1	−185,532 ± 8.3	0	0	371,063.9	1	–	–
	2	−185,331 ± 9.46	52	201.1	370,765.6	0.64	0.36	–
	3	**−185,249 ± 13.47**	**104**	**282.9**	**370,706.1**	**0.58**	**0.26**	**0.16**
Costaceae	1	−28,352.8 ± 10.04	0	0	56,705.6	1	–	–
	2	**−28,141.4 ± 27.52**	**130**	**211.4**	**56,542.8**	**0.94**	**0.06**	**–**
	3	−28,114.7 ± 13.51	260	238.1	56,749.4	0.91	0.06	0.03

w_1, w_2 and w_3 are the mean posterior weights associated with each branch length set.
[a]The number of additional parameter over the model with one branch length set.

To illustrate the information in extra branch length sets we selected data from Rydin et al. (2002) for further analysis. This data set contains two protein coding and two ribosomal sequences and supported two branch length sets. The proportion of the likelihood associated with each branch length set can be calculated on a site-by-site basis. In Fig. 2.3 we plot the proportion associated with the first branch length set, separately for each site in the alignment. Branch length set 1 seems to describe the evolution of rbcL and atpB, whereas the two ribosomal sequences are better described by the second set. The results show a complex pattern of evolution, as not all sites in 26s and 18s are better described by set 1 and some sites favour set 2.

Most of the heterotachy in these data is accounted for by a single clade (Fig. 2.4). The branch length set that favours the protein coding genes (set 1) is longer overall and shows one clade with an accelerated rate compared with the rest of the taxa. The ribosomal genes evolve slowly in this clade, and in general have a more uniform rate throughout the tree (Fig 2.4).

2.8 Discussion

The heterotachy mixture model presented here appears to perform well on simulated data and in the five real data sets. The claim that "… maximum likelihood and BMCMC (Bayesian Markov chain Monte Carlo) can become strongly biased and

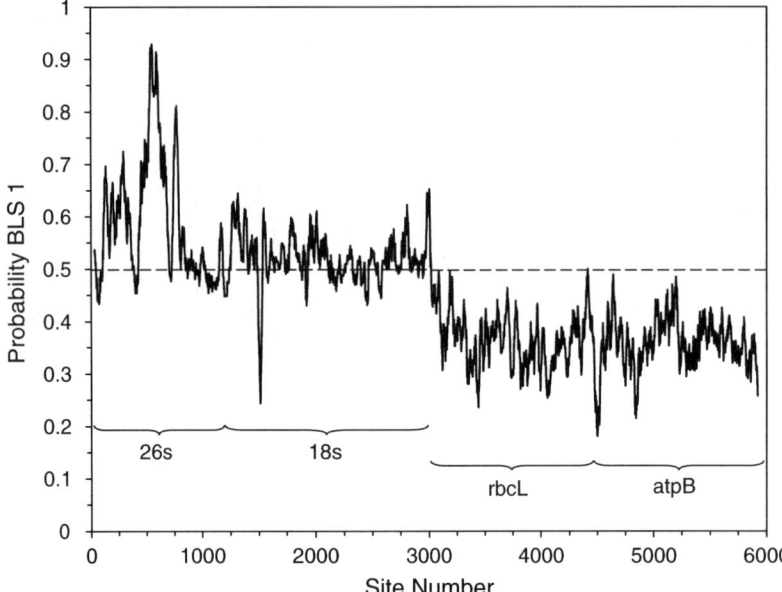

Fig. 2.3 The proportion of likelihood associated with the first branch length set (of two) plotted on a site-by-site basis. The x-axis is the site number, the y-axis is the proportion of likelihood associated with the first branch length set. Value greater than 0.5 are sites that fit the first set better than the second, values less then 0.5 are sites that prefer the second set. The plotted line is a running average over the previous 30 sites, this is used to smooth the line and highlight the signal found in the data. The four genes 26*s*, 18*s*, *rbcL* and *atpB* and their boundaries are depicted by *braces*

statistically inconsistent when the rates at which sequence sites evolve change nonidentically over time." (Kolaczkowski and Thornton 2004) may be a case of 'shooting the messenger'. We show that when mixture models are used to describe data that contain sites whose evolutionary rates vary, the correct topologies are recovered and outperform more conventional models, including parsimony. This identifies the culprit as a misspecified model of evolution, rather than the likelihood method (including MCMC).

Heterotachy may be common in gene-sequence evolution, and may therefore be a common source of error in phylogenetic inference. As our four-taxon simulations show (see also Kolaczkowski and Thornton 2004), heterotachy will mislead models that expect all sites to come from a homogeneous process, whether these models are explicitly defined, such as in likelihood approach, or more implicitly defined, such as in parsimony. Problems associated with heterotachy are likely to increase as the availability of molecular data grows. Studies using hundreds of genes are already being reported (Rokas et al. 2003).

The mixed branch length set model requires the estimation of $2n - 3$ branch lengths and a weighting term for each additional branch length set, where n is the number of taxa. For data sets with a large number of taxa this can represent

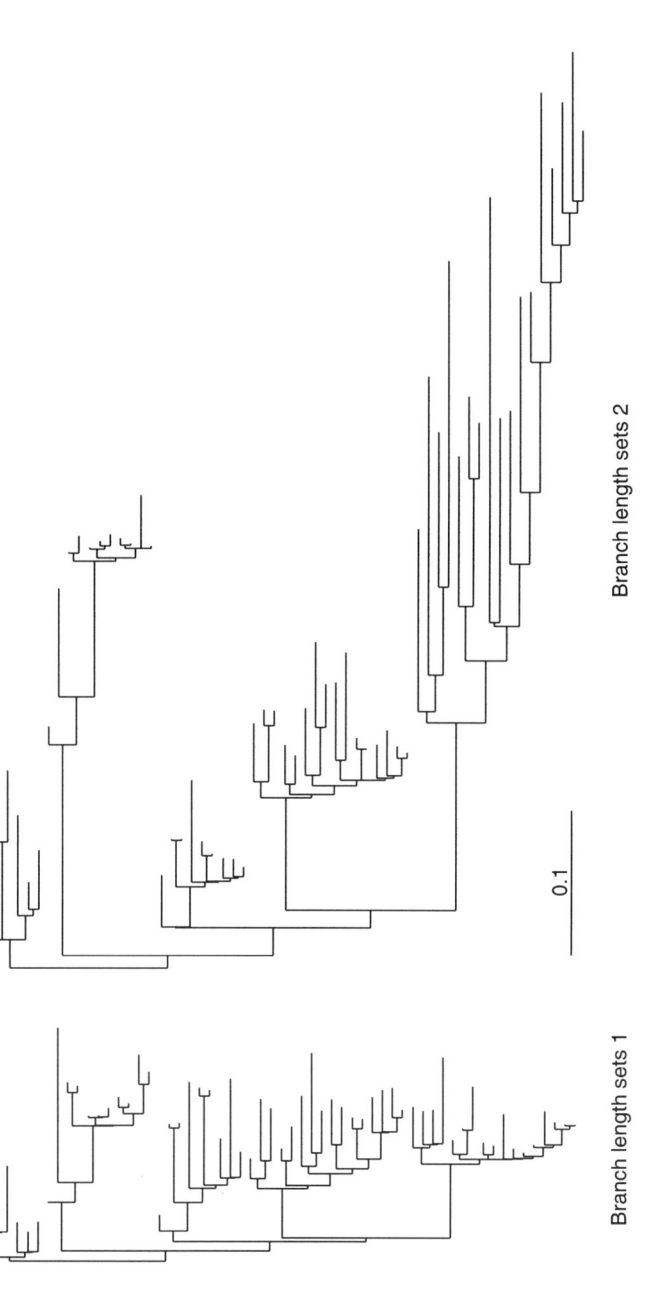

Fig. 2.4 Topology and branch length sets from Rydin et al. (2002). Taxa have been removed for drawing purposes. The topology is the relaxed majority rule consensus, with posterior branch length values used as branch lengths

a substantial number of additional parameters to estimate. The 70-taxon simulations demonstrated that the model is able to detect the correct number of branch length sets and that, in conjunction with the AIC, it does not add branch length sets that are not supported by the data. Small weights assigned to a branch length set and very small likelihood improvements are the signatures of overparameterisation. With real data where the true number of branch sets is not known overparameterisation may be harder to detect. The statistical justification for the addition of an extra set may be based on a small number of branches having a large effect, while the majority have little or no contribution to the likelihood. We are working on a reversible-jump method that may be able to identify which branches in a branch length set are making the important contributions.

The BLS mixture model is implemented in the Bayesian phylogenetic inference package BayesPhylogenies, and is available from the authors' Web site (http://www.evolution.reading.ac.uk). The BLS mixture model is suitable for both nucleotide and morphological data. It can be combined with partitioning or mixture models on substitution matrices, with a number of different substitution models, as well as gamma rate heterogeneity and covarion models.

Acknowledgements This work was supported by grant NE/C51992X/1 from the Natural Environment Research Council, UK, to M.P. Earlier version of the model repeated here were presented at the annual New Zeeland phylogenetics conference (Whitianga, 2005) and the Newton Institute (Cambridge, UK, 2007).

References

Akaike H (1974) A new look at the statistical model identification. IEEE Trans Autom Control 19:716–723
Ane C, Burleigh JG, McMahon MM, Sanderson MJ (2005) Covarion structure in plastid genome evolution: a new statistical test. Mol Biol Evol 22:914–924
Buchheim MA, Michalopulos EA, Buchheim JA (2001) Phylogeny of the Chlorophyceae with special reference to the Sphaeropleales: A study of 18S and 26S rDNA data. Journal of Phycology, 37:819–835
Felsenstein J (2004) Inferring phylogenies. Sinauer, Sunderland
Fitch WM, Markowitz E (1970) An improved method for determining codon variability in a gene and its application to the rate of fixation of mutations in evolution. Biochem Genet 4:579–593
Galtier N (2001) Maximum-likelihood phylogenetic analysis under a covarion-like model. Mol Biol Evol 18:866–873
Gelman A (2003) Bayesian data analysis. CRC, Boca Raton
Geyer CJ (1992) Practical Markov chain Monte Carlo. Stat Sci 7:473–483
Huelsenbeck JP (1999) Variation in the pattern of nucleotide substitution across sites. J Mol Evol 48:86–93
Huelsenbeck JP (2002) Testing a covariotide model of DNA substitution. Mol Biol Evol 19:698–707
Kiontke K, Gavin NP, Raynes Y, Roehrig C, Piano F, Fitch DHA (2004) Caenorhabditis phylogeny predicts convergence of hermaphroditism and extensive intron loss. Proceedings of the National Academy of Sciences of the United States of America, 101:9003–9008

Kolaczkowski B, Thornton J (2004) Performance of maximum parsimony and likelihood phylogenetics when evolution is heterogeneous. Nature 431:980–984

Koshi JM, Goldstein RA (1998) Models of natural mutations including site heterogeneity. Proteins Struct Funct Genet 32:289–295

Lopez P, Casane D, Philippe H (2002) Heterotachy, an important process of protein evolution. Mol Biol Evol 19:1–7

Mueller RL, Macey JR, Jaekel M, Wake DB, Boore JL (2004) Morphological homoplasy, life history evolution, and historical biogeography of plethodontid salamanders inferred from complete mitochondrial genomes. Proceedings of the National Academy of Sciences of the United States of America, 101:13820–13825

Pagel M, Meade A (2004) A phylogenetic mixture model for detecting pattern-heterogeneity in gene sequence or character-state data. Syst Biol 53:571–581

Pagel M, Meade A (2005) Mixture models in phylogenetic inference. In: Gascuel O (ed) Mathematics of evolution and phylogeny. Clarendon, Oxford, pp 121–142

Philippe H, Zhou Y, Brinkmann H, Rodrigue N, Delsuc F (2005) Heterotachy and long-branch attraction in phylogenetics. BMC Evol Biol 5:50

Piel WH, Donoghue MJ, Sanderson MJ (2002) TreeBASE: a database of phylogenetic knowledge. pp 41–47 in: J. Shimura, K. Wilson, and D. Gordon, eds. The interoperable "Catalog of Life." Research Report, National Institute for Environmental Studies No. 171, Tsukuba, Japan

Rambaut A, Grassly NC (1997) Seq-Gen: an application for the Monte Carlo simulation of DNA sequence evolution along phylogenetic trees. Comput Appl Biosci 235–238

Rokas A, Williams BL, King N, Carroll SB (2003) Genome-scale approaches to resolving incongruence in molecular phylogenies. Nature 425:798–804

Rydin C, Kallersjo M, Friist EM (2002) Seed plant relationships and the systematic position of Gnetales based on nuclear and chloroplast DNA: Conflicting data, rooting problems, and the monophyly of conifers. International Journal of Plant Sciences, 163:197–214

Shalchian-Tabrizi K, Skanseng M, Ronquist F, Klaveness D, Bachvaroff TR, Delwiche CF, Botnen A, Tengs T, Jakobsen KS (2006) Heterotachy processes in rhodophyte-derived secondhand plastid genes: implications for addressing the origin and evolution of dinoflagellate plastids. Mol Biol Evol 23:1504–1515

Specht CD (2006) Systematics and evolution of the tropical monocot family Costaceae (Zingiberales): A multiple dataset approach. Systematic Botany, 31:89–106

Spencer M, Susko E, Roger AJ (2005) Likelihood, parsimony, and heterogeneous evolution. Mol Biol Evol 22:1161–1164

Swofford DL, Olsen GJ, Waddell PJ, Hillis DM, Moritz C, Mable BK (1996) Mol Syst 407–514

Taylor MS, Kai C, Kawai J, Carninci P, Hayashizaki Y, Semple CA (2006) Heterotachy in mammalian promoter evolution. PLoS Genet 2:e30

Tuffley C, Steel M (1998) Modeling the covarion hypothesis of nucleotide substitution. Math Biosci 147:63–91

Venditti C, Meade A, Pagel M (2008) Phylogenetic mixture models can reduce the node-density artifact. Syst Biol, in press 2008

Yang Z (1994) Maximum likelihood phylogenetic estimation from DNA sequences with variable rates over sites: approximate methods. J Mol Evol 39:306–314

Part II
Concepts in Evolutionary Biology

Chapter 3
Accelerated Evolution of Genes of Recent Origin

Macarena Toll-Riera, Jose Castresana, and M. Mar Albà

Abstract The gene content of any genome is a rich mosaic of genes that have originated at different times during evolution. Among the most interesting properties related to gene age is the fact that younger genes tend to show accelerated evolutionary rates with respect to older genes. Here, we use a large number of closely related mammalian genomes to gain further insights into the relationship between gene age and evolutionary rate. We define a group of primate-specific genes that are absent from 11 non-primate mammalian genomes as well as from other eukaryotic genomes. These genes, of very recent origin, show the highest evolutionary rate and the shortest protein length. We discuss how these results may shed light on understanding the proposed mechanisms for the origin of lineage-specific, novel genes.

3.1 Introduction

Genes are in a continuous state of change owing to mutational and selective processes. The rate of change experienced by genes during their evolution can be quantified by comparing the gene sequences of contemporaneous species. In coding sequences, nucleotide substitution rates can be synonymous (K_s), when they do not

---

M. Toll-Riera and M.M. Albà
Research Unit on Biomedical Informatics, Institut Municipal d'Investigació Mèdica, Universitat Pompeu Fabra, E-08003 Barcelona, Spain
mtoll@imim.es

J. Castresana
Institute of Molecular Biology of Barcelona, CSIC, 08034 Barcelona, Spain
jcvagr@ibmb.csic.es

M.M. Albà
Catalan Institution for Research and Advanced Studies, 08010 Barcelona, Spain
malba@imim.es

result in amino acid changes, or non-synonymous (Ka), when they result in such changes. Ks, therefore, reflects the background gene mutation level, whereas Ka reflects selection at the protein level. Low Ka/Ks values indicate that there exists a strong negative selection on the protein sequence, in other words, that the protein is subject to important functional constraints. High Ka/Ks values, instead, indicate relaxed selective constraints and/or positive selection due to the frequent occurrence of advantageous mutations.

At present there are various mammalian genome sequences available for comparative studies, which allows new investigations to be undertaken on the mechanisms of gene evolution and their functional implications. Previous studies have revealed that, among the genes of any genome, there is a great disparity in the Ka/Ks values measured using alignments of orthologous genes. Different factors have been shown to correlate or influence gene evolutionary rate: gene function, expression breadth and intensity, essentiality of the genes for the survival of the organism, number of protein–protein interactions, and taxonomic distribution, among others (Hurst and Smith 1999; Albà and Castresana 2005; Furney et al. 2006; Pal et al. 2006).

In any particular genome, some of the genes have homologues in many other organisms, while others have a highly restricted distribution, with homologues in only a few other species. Novel genes that have no recognizable homologues in other species, or that have homologues only in very closely related species, have been named orphan genes. These genes are of special interest as they can play important roles in lineage-specific adaptative processes. For example, it has been reported that around 26–29% of *Drosophila* proteins do not show any significant sequence similarity match with non-insect sequences (Domazet-Loso and Tautz 2003). Interestingly, these *Drosophila* orphan genes evolve much faster than non-orphan genes (Domazet-Loso and Tautz 2003). Another study has identified 154 groups of insect-specific proteins, with homologues in *Drosophila melanogaster*, *Anopheles gambiae*, *Bombyx mori*, *Tribolium castaneum* and *Apis mellifera* but not in *Homo sapiens*, *Caenorhabditis elegans*, and *S Saccharomyces cerevisiae* (Zhang et al. 2007). These genes are enriched in stress- and stimulus-response proteins, together with cuticle and pheromone/odorant binding proteins. Again, proteins that are specific of the *Drosophila* lineage in this study show increased Ka/Ks ratios with respect to proteins with a wider taxonomic distribution. Studies in bacterial genomes have also revealed the existence of a large number of genes that are specific of a species or taxon of closely related organisms, and which tend to evolve quickly (Daubin and Ochman 2004). In the analysis of the complete mouse genome sequence (Waterston et al. 2002) it was found that about 14% of the mouse genes were mammalian-specific (with no homologues in other chordates), 4% chordate-specific (with no homologues in other metazoans), 27% metazoan-specific (with no homologues in other eukaryotes), 29% eukaryote-specific (with no homologues in prokaryotes) and 23% universal (with homologues both in eukaryotes in prokaryotes). A recent analysis has classified *Drosophila melanogaster* genes in different levels according to their evolutionary origin, as inferred from BLAST sequence similarity searches in extant species (Domazet-Loso et al. 2007). According to this study, 36.4% of the genes would have originated in the last common ancestor of all cellular organisms, 7.6% in the Eumetazoa ancestor and as many as 17.6% in the Diptera ancestor.

An important property of genes that have different age or taxonomic distribution is that the younger the group considered the higher its average evolutionary rate (Ka or Ka/Ks). This has been explicitly quantified in two recent papers. First, Albà and Castresana (2005) described an inverse correlation between gene age and evolutionary rate in mammalian genes. Later, Cai et al. (2006) found a very similar relationship between lineage specificity and gene evolutionary rate in fungi. The effect of lineage specificity (gene age) was found to be stronger than the influence of other factors previously proposed to be predictors of a gene's evolutionary rate, including expression level, gene essentiality or number of protein–protein interactions (Cai et al. 2006). All these results indicates, a priori, that there exist strong differences in the evolutionary capacity of genes that represent recent innovations with respect to genes that have ancestral functions and experience stronger sequence constraints.

In the study of Albà and Castresana (2005), Ka and Ks values were estimated from human and mouse orthologous pairs. Similarly to previous studies (Lander et al. 2001; Waterston et al. 2002), the genes were classified in different age groups depending on the presence or absence of homologues in other available genomes using BLASTP searches (E-value less than 10^{-4}). Six different genomes were employed: *Takifugu rubripes*, *Drosophila melanogaster*, *Caenorhabditis elegans*, *A Arabidopsis thaliana*, *Saccharomyces cerevisiae* and *Schizosaccharomyces pombe*. Genes present in all these eukaryotic genomes were classified as "old", those present in *Takifugu rubripes*, *Drosophila melanogaster* and *Caenorhabditis elegans* but absent from the other genomes as "metazoans", those present in *Takifugu rubripes* but absent from the other genomes as "deuterostomes" and genes absent from the six genomes as "tetrapods". The youngest group, "tetrapods", evolved, on average, almost 4 times more rapidly than the "old" group.

In the study mentioned above, the most novel group was that of "tetrapods", composed of genes only found in human and mouse. These genes, for which homologues in *Takifugu rubripes* could not be found, may have originated, approximately, from 400 million to 90 million years ago. What is the behaviour of other, more recent, mammalian genes? To investigate this issue here, we identify primate-specific genes and compare their evolutionary rates, as well as other gene properties, with those of groups of older genes. In addition, we use gene datasets from a much larger number of complete genome sequences, 27, to increase the resolution of the age groups. In the last section of this work, we also describe possible problems in the identification of homologues of fast-evolving genes. This study reinforces the previous observations of an inverse relationship between gene age and evolutionary rate. Furthermore, the data presented here pose new questions in relation to the change in constraints that genes may experience during their history and the origin of new genes.

3.2 Results

3.2.1 Definition of Groups of Genes of Different Age

Human and macaque orthologous protein pairs, their corresponding gene coding sequences, chromosome and Gene Ontology annotations were extracted from the Ensembl database using the Biomart datamining tool (Flicek et al. 2007). In Ensembl, orthologous genes are defined on the basis of maximum reciprocal similarity, conserved synteny and phylogenetic tree reconstruction. When more than one protein sequence per gene was available we chose the longest one. We obtained 19,352 human–macaque orthologues. In order to build groups of primate genes of different age or taxonomic distribution, we obtained protein sequences from 25 additional eukaryotic genomes: 12 mammals (*Pan troglodytes, Mus musculus, Rattus norvergicus, Bos taurus, Canis familiaris, Echinops telfari, Loxodonta africana, Felis catus, Dasypus novemcinctus, Oryctolagus cuniculus, Monodephis domestica* and *Ornithorhynchus anatinus*), five non-mammalian vertebrates (*Gallus gallus, Xenopus tropicalis, Danio rerio, Tetraodon nigroviridis* and *Takifugu rubripes*), two urochordates (*Ciona intestinalis* and *Ciona savigny*), two insects (*Drosophila melanogaster* and *Anopheles gambiae*), one nematode (*Caenorhabditis elegans*), two fungi (*Saccharomyces cerevisiae* and *Schizosaccharomyces pombe*) and one plant (*Arabidopsis thaliana*). All sequences were downloaded from Ensembl, except for those of *Saccharomyces cerevisiae, Schizosaccharomyces pombe, Arabidopsis thaliana, Caenorhabditis elegans, Drosophila melanogaster* and *Takifugu rubripes*, which were obtained from the Cogent Database release 153 (Goldovsky et al. 2005).

We classified the 19,352 human–macaque orthologues in four age groups—Primates, Mammals, Vertebrates and Eukarya old—using BLASTP sequence similarity searches (Altschul et al. 1997). We considered that there existed a homologue in another genome if there was at least one BLASTP hit with an expectation value (E-value) smaller than 10^{-4}. To avoid the interference caused by possible low-complexity sequences we filtered this type of region from the human sequences using the SEG program (Wootton and Federhen 1996). If the human protein had any homologue in *Pan troglodytes* and *Macaca mulatta*, but not in the rest of genomes, it was classified as Primates. If the protein had homologues in the other 11 mammalian species, but not in the rest of eukaryotes, it was classified as Mammals. If it had homologues in all the species mentioned before, and also in all other vertebrates tested, but not in the rest of the eukaryotes, it was classified as Vertebrates. Finally, if it had homologues in all 25 eukaryotes tested, it was classified as Eukarya old (Fig. 3.1).

3.2.2 Evolutionary Rate of Genes of Different Age

Once the groups had been defined we measured evolutionary rate and protein length distribution in each group and performed across-group comparisons. In order to

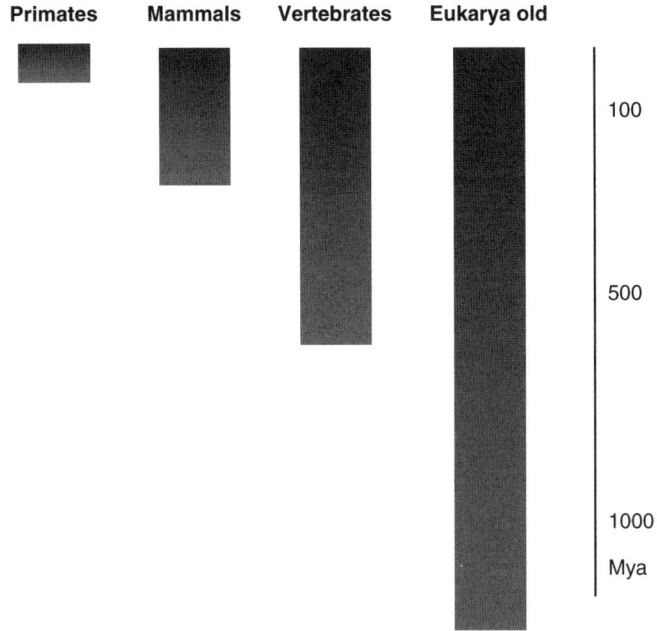

Fig. 3.1 Evolutionary lifespan of human genes classified in different age groups. Gene age is defined by the lack of homologues in other eukaryotic genomes. *Mya* million years ago

measure non-synonymous (Ka) and synonymous (Ks) substitution rates we used coding sequence alignments for each human and macaque protein orthologous pair. The coding sequence alignments were based on protein alignments, which were obtained by ClustalW (Thompson et al. 1994). Ka and Ks were estimated using the maximum-likelihood method implemented in the Codeml program of the PAML software package (Yang 2007). Pairs with high substitution rates (Ka > 0.3 and/or Ks > 2 substitutions per positions) were discarded to avoid including not bona fide orthologues in the analysis. After applying this filter, we obtained a dataset that contained 2.610 gene pairs: 2.063 classified as Eukarya old, 373 as Vertebrates, 109 as Mammals and 65 as Primates.

We found that Ka and Ka/Ks values were significantly different in all group-to-group comparisons ($p < 0.01$, using the Kolmogorov–Smirnov test), except in the case of the differences between Primates Ka and Mammals Ka, which were not significant. In Fig. 3.2 it can be observed that the more recent the age group, the more elevated the Ka/Ks values. Average Ka values were 0.0917 substitutions per position for Primates, 0.0619 for Mammals, 0.0351 for Vertebrates and 0.0219 for Eukarya old. Average Ka/Ks values were 1.0337 for Primates, 0.7126 for Mammals, 0.354 for Vertebrates and 0.205 for Eukarya old. This fits the average Ka/Ks value (0.247) previously obtained for a large collection of human and macaque orthologues (Gibbs et al. 2007). The median values were always lower than the mean values, reflecting the non-normal shape of the distribution and the excess of proteins

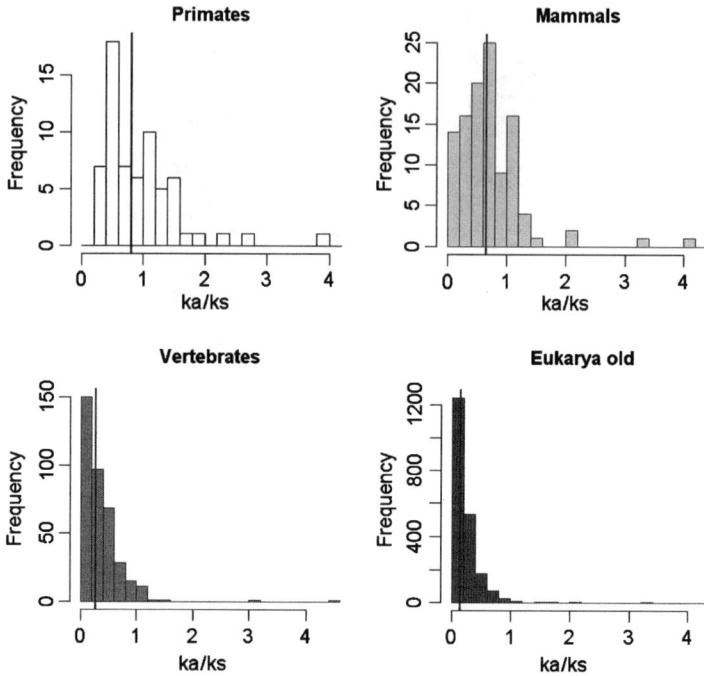

Fig. 3.2 Distribution of Ka/Ks in different gene age groups. The *vertical black line* marks the median

with evolutionary rates lower than the average (Fig. 3.2). For example, in the case of primate-specific proteins the average Ka/Ks was 1.0337, whereas the median was 0.84, indicating that the majority of the proteins had Ka/Ks lower than 1. So, as in Albà and Castresana (2005), we found an accelerated evolutionary rate in younger genes with respect to more ancient ones. In contrast to the previous study, here we included genes of very recent origin, only found in the primate genomes (primate orphan genes). A number of these genes showed Ka/Ks larger than 1, indicating that they may be subject to positive selection. Interestingly, an important fraction of the mammalian genes, with homologues in 13 other mammalian genomes besides human, also showed Ka/Ks larger than 1 (Fig. 3.2). These genes are likely to play important functions in the context of lineage-specific adaptations.

In comparison with the study of Albà and Castresana (2005), the non-synonymous to synonymous substitution rate ratios (Ka/Ks) measured here for equivalent groups showed higher values. For example, for the oldest group (Eukarya old) the average Ka/Ks is 0.205 in the present study, which used human and macaque orthologues, whereas it was 0.082 in the other study, using human and mouse orthologues. This agrees with the recent observation that primate Ka/Ks rates are comparatively higher than rodent Ka/Ks rates. It has been proposed that such differences may be due to a reduction in purifying selection in hominids or a consequence of the smaller hominid population size (Gibbs et al. 2007).

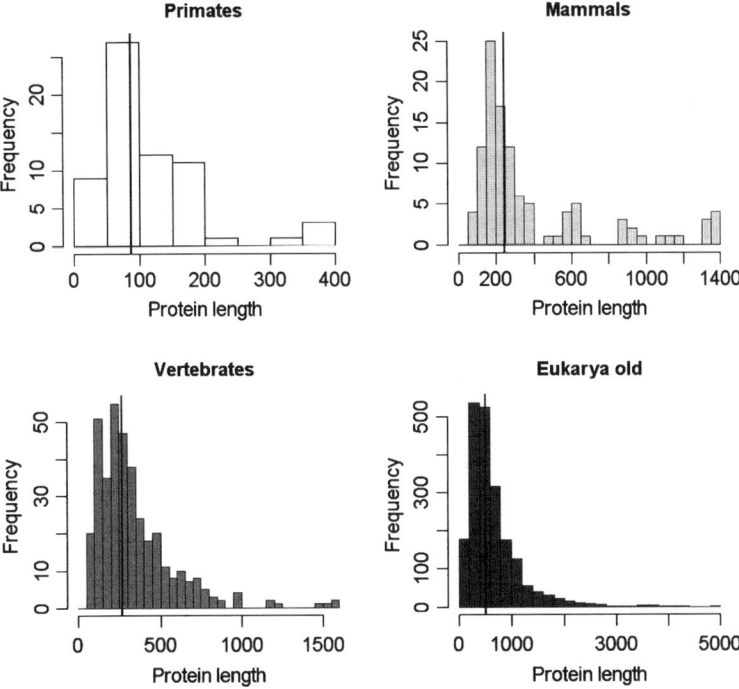

Fig. 3.3 Distribution of protein length in different gene age groups. The *vertical black line* marks the median

3.2.3 Length of Genes of Different Age

Next, we measured the protein length of genes classified in the different age groups. We found a direct relationship between the age of the gene and the length of the encoded protein (Fig. 3.3). The most novel genes (Primates) were the shortest ones (average 114.05 and median 86 residues), when compared with Mammals (average 390.29 and median 243), Vertebrates (average 340.7 and median 270) and Eukarya old (average 652.9 and median 517). In addition, an important fraction of primate-specific genes, 49%, only had one exon. No significant differences between groups were found in GC content, indels and fraction of the protein occupied by low-complexity segments. The very short length of primate orphan genes was striking. However, orphan genes in other species have also been reported to be particularly short (Domazet-Loso and Tautz 2003; Levine et al. 2006).

3.2.4 Functions of Primate Genes of Different Age

We obtained Gene Ontology (GO) annotations for genes classified in different groups from Ensembl. These annotations included terms found in the three GO ontologies "cellular compartment", "biological process" and "molecular function"

Table 3.1 Most commonly found Gene Ontology functions in genes of different age

Primates	Mammals	Vertebrates	Eukarya old
Nucleus	Membrane	Membrane	Nucleus
DNA binding	Extracellular region	Immune response	Intracellular
Regulation of transcription, DNA-dependent	Protein binding	Protein binding	Protein binding
Negative regulation of progression through cell cycle	Nucleus	Receptor activity	Zinc ion binding
Integral to membrane	Immune response	Signal transduction	Nucleic acid binding
Cell cycle	Intracellular	Cytoplasm	Metal ion binding

(Harris et al. 2004). Differences in function distribution are apparent from the list of the top six most commonly-found GO annotations (Table 3.1). For example, in Mammals and Vertebrates the most frequently found annotation is "membrane", a term that does not appear in the top list of Eukarya old gene GO annotations. A similarly biased term is "immune response", present in the top list of Mammals and Vertebrates but absent from the top list of functions of older genes. In the group Primates we found an enrichment of "DNA-binding" and "regulation of transcription" terms. Some of these annotations, however, are only based on protein domain predictions. The Primate group was much more poorly annotated than the rest of the groups, with only 20% of genes being associated to at least one GO term. In contrast, this figure was 67.9% for Mammals, 81.23% for Vertebrates and 96.17% for Eukarya old. So, primate-specific genes remain largely uncharacterized.

Among the small number of primate orphan genes with known function we identified a number of genes related to the immune response. One example was the minor histocompatibility antigen HB-1 gene. This gene encodes one of the minor histocompatibility antigens, which plays an important role in the induction of cytotoxic T lymphocyte reactivity against leukaemia. This gene is only expressed in B cell acute lymphoblastic leukaemia cells and Epstein–Barr virus transformed B cells (Dolstra et al. 1999). Another immune response related gene was dermcidin precursor. This gene encodes a secreted protein that is subsequently processed into mature peptides of distinct biological activities. The C-terminal peptide is constitutively expressed in sweat and has antibacterial and antifungal activities. Antimicrobial peptides are an important component of the innate response. The dermcidin peptide shows antimicrobial activity against a wide range of pathogenic microorganisms. It has been hypothesized that sweat is implicated in the regulation of human skin flora through this peptide, and it may help to limit the infection by pathogens during the first hours following bacterial colonization (Schittek et al. 2001).

Other primate-specific genes were expressed in tumours. Chondrosarcoma-associated protein 2 was annotated with the GO term "response to drug". This protein is expressed in ovary, kidney and in various tumour cell lines, and it is associated with the chemotherapy-resistant and neoplastic phenotype (Duan et al. 1999). In the group Primates we also identified several members of the SPAN-X protein family. These proteins are expressed in testis and sperm and may play a role in spermatogenesis. They are also cancer-testis antigens (Kouprina et al. 2004).

3.2.5 Estimation of the Age of Genes

The estimation of the age of genes by sequence similarity searches may be problematic, especially for rapidly evolving proteins and evolutionary distant comparisons. In such cases, the similarity may be no longer detectable, and the genes may be erroneously classified in younger groups. When does this become a relevant problem? The group of primate-specific genes is defined by the lack of homologues in other closely related species (mammals), and therefore they are very unlikely to be affected by problems in BLASTP sensitivity. Mammalian-specific genes, however, contain a significant fraction of fast-evolving genes, and they are defined by lack of homologues in more distantly related species. Albà and Castresana (2005) performed several controls to explore the possibility that BLAST problems to detect homologues would bias the results. The results, however, appeared to be robust. For example, restricting the analysis to a set of highly conserved proteins, for which homologues should be readily detectable, produced the same relationship between gene age and evolutionary rate. However, Elhaik et al. (2006) stated that the results found in Albà and Castresana (2005) were in fact due to lack of sensitivity to detect homologues of fast-evolving proteins. They based their argument on simulations of homogeneously evolving DNA sequences along a phylogenetic tree and subsequent nucleotide BLAST searches to classify sequences in different age groups. In this setup, fast-evolving sequences often failed to find homologues, which resulted in younger groups being highly enriched in rapidly evolving sequences. However, their observations could not really prove that the results in Albà and Castresana (2005) were an artefact, as Albà and Castresana's study was based on protein sequences and not on neutrally evolving DNA sequences. Protein sequences are subject to selection and therefore evolve in a much more constrained manner. In protein BLAST (BLASTP), clusters of functional conserved sites are often sufficient to detect homology, even if the rest of the sequence is poorly conserved. To clarify this controversy, Albà and Castresana (2007) tested whether the different age groups that they had previously described could have arisen by a protein BLAST artefact. To do so, they simulated the evolution of protein sequences along a phylogenetic tree with branch lengths proportional to the times separating the previously used genomes, using the program Rose (Stoye et al. 1998). The tree was multiplied by different factors in order to obtain sequences of overall different evolutionary rates. In addition, as these were protein sequences, they were simulated using among-site rate heterogeneity, which accounts for the fact that in real proteins some positions will be more conserved than others. Spatial patterns of rate heterogeneity were extracted from a subset of eukaryotic protein alignments using TreePuzzle (Schmidt et al. 2002). Once the simulated sequences along the trees had been generated, a set of genes with evolutionary rates equivalent to those observed in human and mouse orthologues was randomly pooled. The equivalent to the human sequence, in the simulated dataset, was then classified as "old", "metazoans", "deuterostomes" and "tetrapods" using BLASTP sequence similarity searches against the other sequences. As all the sequences had been evolved along the complete tree, those that were not classified as "old" were in fact misclassified owing to failure of BLASTP to detect the homologues. The total number of misclassified genes in the simulation study was very small, leading to the conclusion that the accelerated rate of evolution of young genes with respect to older genes was not a BLAST artefact (Albà and Castresana 2007).

In summary, these studies showed that mammalian novel genes, even if rapidly evolving, should be, in the vast majority of cases, detectable in other eukaryotic genomes by BLASTP, if they were present in such genomes. In the case of the primate-specific genes presented here, homologues in other mammalian genomes should be readily detectable if they existed, given their high evolutionary proximity. Then, novel genes are not just highly diverged genes but need to have originated recently by some other mechanism.

3.3 Discussion

3.3.1 Differences between Genes of Different Age

As seen in this study, and in many others (Lander et al. 2001; Waterston et al. 2002; Cai et al. 2006; Domazet-Loso et al. 2007; Zhang et al. 2007), genomes contain a mosaic of genes formed at different time points during evolution. Here, using a large number of closely related genomes, we have observed that younger primate proteins evolve more rapidly than older ones, reinforcing previous results on the existence of an inverse relationship between age and evolutionary rate in mammalian genes (Albà and Castresana 2005). Most of the studies that try to date the age of genes use BLAST sequence similarity searches to determine the presence or absence of homologues in other organisms. This type of search should produce reliable results for closely related organisms (for example within mammals or vertebrates) but less reliable ones for very distant taxa. Nevertheless, studies based on protein structural domains, where homologous structures are grouped together even in the absence of significant sequence similarity, result in a similar mosaic picture, where some structural folds are older than others (Abeln and Deane 2005; Choi and Kim 2006). Interestingly, structures of different age show different properties. Ancient structures are generally long and enriched in the α/β class, structures of intermediate age tend to be of moderate length and are α, β, $\alpha + \beta$ or α/β, and the youngest structures are shorter and typically α, β or $\alpha + \beta$ (Choi and Kim 2006). Following these observations, the authors have proposed that proteins tend to evolve from α or β small structures, present in the initial phases, to larger, α/β, structures when they become "mature" proteins. Therefore, proteins may experience important changes along their evolutionary history and these changes would have some kind of directionality: proteins would become gradually longer and more complex in their structure, until they would presumably reach a state of equilibrium.

In a similar manner, the observation that older proteins show slower evolutionary rates than younger proteins raises the interesting question of whether proteins tend to experience a decrease in their speed of evolution as they become older, as hypothesized by Albà and Castresana (2005). This could be explained by an increase in the strength of purifying selection with time, perhaps linked to the acquisition of additional functions and functional fine-tuning. Present-day proteins of different age would, in this view, belong to different stages of protein evolution. Alternatively, it could be that the characteristics of proteins born at different times in life's history would be the main determinant of their typical speed of evolution. As seen before, some functions are more typical of young proteins than of old proteins, and it has been known for a long time that proteins associated with different functions evolve at typically different rates. An analysis by GO functional class, however, did show that the same pattern of inverse relationship between gene age and evolutionary rate could be observed within each class (Albà and Castresana 2005). For example, RNA-binding proteins, which are among the slowest-evolving protein functional classes, show higher evolutionary rates in younger groups than in older groups. The existence of this particular type of non-clock model of evolution for the youngest genes could be tested using evolutionary tree reconstruction methods allowing branch length variation (Seo et al. 2004; Yang 2007). In young genes, increased evolutionary rates (branch length) should be observed at the base of the tree, near the origin of the gene. In contrast, older genes should show a more homogeneous evolution along the tree. Some results seem to point to such a non-homogeneous mode of evolution for genes of recent origin (Iwabe et al. 1996; Miyata and Suga 2001).

The formation of new genes is an important source of functional innovation and thus the study of these genes is crucial to understand how organisms acquire novel biological functions and adapt to new conditions of life. As shown in this study and others (Martinez-Morales et al. 2007), genes restricted to mammals or vertebrates are enriched in extracellular, immune response and receptor functions. The diversification of genes involved in signalling, cell communication and response to external elements has thus accompanied vertebrate evolution. In contrast, genes present in all eukaryotes are enriched in intracellular and metabolic functions. The study of the age of genes involved in the formation of the neural crest, a vertebrate-specific physiological process, revealed that 9% of genes are vertebrate innovations (Martinez-Morales et al. 2007). This illustrates how the evolution of a novel developmental process takes into play the recruitment of existing proteins and regulatory networks together with the emergence of new functional genes in the genome.

3.3.2 Properties of Novel Genes

In this study, we have included a group of genes of very recent origin, only present in primate genomes and which may have originated between 60 million and 30 million years ago. These primate-specific proteins are short, show evidence of positive selection and, in most cases, do not have a known function. Orphan genes are typically short in all other organisms studied (Domazet-Loso and Tautz 2003; Daubin and Ochman 2004; Clark et al. 2007). Short proteins often go undetected by gene prediction programs and therefore are likely to be underrepresented in protein databases. Interestingly, a recent estimation of the proportion of short proteins (fewer than 100 amino acids) in the mouse genome, based on new complementary DNA data and experimental validation of some of the predicted peptides, has elevated the estimated number of short proteins from about 3% to about 10% of the proteome (Frith et al. 2006). However, it may be that some predicted short proteins in transcript databases are spurious open reading frames that are not functional (Clamp et al. 2007). Large-scale experimental protein validation studies will be needed to determine the real number of short functional orphan proteins.

Novel genes are characterized by a fast pace of evolution, denoting weak negative selection. On the other hand, positive selection is also expected to be a relevant evolutionary force in young genes, as the gene is, presumably, in the process of adapting to a new function. For example, the primate-specific family *morpheus* has experienced high levels of amino acid replacement during its evolution, suggesting adaptive evolution during the emergence of humans and African apes (Johnson et al. 2001). Several novel genes described in *Drosophila* also exhibit patterns of non-synonymous and synonymous substitutions that suggest positive selection (Levine et al. 2006). In the study presented here, an important number of primate- and mammalian-specific genes show non-synonymous to synonymous substitution rates ratio (Ka/Ks) larger than 1, compatible with positive selection acting on a significant number of amino acids along the sequence.

3.3.3 Hypotheses to Explain the Origin of Novel Genes

The presence of a significant number of orphan genes in any particular genome is now well established, but how these genes originated in the first place remains enigmatic. Understanding the origin of novel genes is crucial in evolutionary genomics but, like understanding the origin of novel organs during the evolution of life, it is not an easy task. First, we have to consider that, during evolution, everything (genes, organs, etc.) evolves from previous structures. On the other hand, punctuated evolution and fast adaptation processes may break the link of homology with previous structures and it is then when we consider that novel genes or novel organs arise. The existence of a high number of novel genes in different lineages was discovered thanks to the analysis of the complete genomes of a growing number of organisms. Since then, several hypotheses have been proposed to try to explain how these genes originated. It is likely that no single mechanism can explain the origin of all novel genes and that different genes arose by different mechanisms.

Following a study of orphan genes in *Drosophila*, it was proposed that one mechanism of gene birth could be gene duplication followed by rapid divergence (Domazet-Loso and Tautz 2003). In this model, after a gene duplication, one copy would go through a phase of very rapid evolution due to relaxed selective constraints (Lynch and Conery 2000). This would be followed by fast adaptive evolution due to the acquisition of a new function. Finally, the rate of change would slow down owing to the effect of negative selection, but the sequence would now be too different to be recognized as homologous to the other copy. The initial phase of very fast evolution would break the homology link with the original gene. However, it is not unlikely that remote homology detection methods such as those based on the protein structure may allow the detection of some of the founding genes.

Another mechanism that has been proposed for the generation of completely novel proteins is frameshift mutation (Ohno 1984). In particular, the release of the pressure to maintain the ancestral function in one of the gene copies after gene duplication can facilitate the formation of essentially new proteins by frameshift events, and indeed, a significant number of such cases have recently been identified (Okamura et al. 2006).

Finally, some genes could also originate directly from non-coding sequences. For example, in notothenioid fish an intronic sequence from an ancestral trypsinogen gene has been recruited into protein coding function in a descendant antifreeze protein (Chen et al. 1997). The newly evolved testis-specific Cdic gene from *Drosophila melanogaster* also seems to have originated from an intronic sequence (Nurminsky et al. 1998). More recent studies suggest that several *Drosophila*-specific testis-expressed genes are derived from ancestral non-coding genome sequences (Begun et al. 2006; Levine et al. 2006). It has been recently discovered that a much larger fraction of the genome than previously thought is expressed (Birney et al. 2007). Thus, it is possible that the occasional expression of transcripts, containing open reading frames, could lead to the production of new peptides, which could then be "tested" by natural selection. If they were advantageous, they would tend to be retained. At later stages peptide sequence, expression level and expression regulation would continue to be shaped by selection. As it is very unlikely that very long open reading frames are formed in non-coding parts of the genome, such new proteins would be expected to be short (60–100 amino acids). Indeed, short protein size is truly characteristic of orphan genes. Furthermore, at least among primate-specific genes, a large proportion of new genes are encoded by a single exon. Given that the length of the protein shows a direct relationship with age, proteins that are initially small may increase their length with time, a process that may be accompanied by added structural complexity (more α/β structures) and a general increase in evolutionary constraints (lower Ka/Ks ratio)

3.4 Conclusion

We have demonstrated that in the primate lineage there is an inverse relationship between age and evolutionary rate, as previously observed in fungi, *Drosophila* and mammals. We have also observed a direct relationship between protein length and gene age. Primate orphan genes are poorly characterized, but some of them have functions related to immune response and male reproduction. How these genes have originated is still not clear, but possible mechanisms include gene duplication, frameshift mutation and birth from non-coding regions. More studies on the function and expression of primates orphan genes will be needed to better understand the specific adaptations that have taken place along the primate lineage.

Acknowledgements We acknowledge funding from Plan Nacional de I + D Ministerio de Educación y Ciencia (BFU2006-07120), AGAUR-Generalitat de Catalunya (M.T-R.) and Fundació ICREA (M.M.A.).

References

Abeln S, Deane CM (2005) Fold usage on genomes and protein fold evolution. Proteins 60:690–700
Albà MM, Castresana J (2005) Inverse relationship between evolutionary rate and age of mammalian genes. Mol Biol Evol 22:598–606
Albà MM, Castresana J (2007) On homology searches by protein Blast and the characterization of the age of genes. BMC Evol Biol 7:53
Altschul SF et al (1997) Gapped BLAST and PSI-BLAST: a new generation of protein database search programs. Nucleic Acids Res 25:3389–3402
Begun DJ, Lindfors HA, Thompson ME, Holloway AK (2006) Recently evolved genes identified from Drosophila yakuba and D. erecta accessory gland expressed sequence tags. Genetics 172:1675–1681
Birney E et al (2007) Identification and analysis of functional elements in 1% of the human genome by the ENCODE pilot project. Nature 447:799–816
Cai JJ, Woo PC, Lau SK, Smith DK, Yuen KY (2006) Accelerated evolutionary rate may be responsible for the emergence of lineage-specific genes in ascomycota. J Mol Evol 63:1–11
Chen L, DeVries AL, Cheng CH (1997) Evolution of antifreeze glycoprotein gene from a trypsinogen gene in Antarctic notothenioid fish. Proc Natl Acad Sci USA 94:3811–3816
Choi IG, Kim SH (2006) Evolution of protein structural classes and protein sequence families. Proc Natl Acad Sci USA 103:14056–14061
Clamp M et al (2007) Distinguishing protein-coding and noncoding genes in the human genome. Proc Natl Acad Sci USA 04:19428–33
Clark AG et al (2007) Evolution of genes and genomes on the Drosophila phylogeny. Nature 450:203–218
Daubin V, Ochman H (2004) Bacterial genomes as new gene homes: the genealogy of ORFans in E. coli. Genome Res 14:1036–1042
Dolstra H et al (1999) A human minor histocompatibility antigen specific for B cell acute lymphoblastic leukemia. J Exp Med 189:301–308
Domazet-Loso T, Tautz D (2003) An evolutionary analysis of orphan genes in Drosophila. Genome Res 13:2213–2219
Domazet-Loso T, Brajkovic J, Tautz D (2007) A phylostratigraphy approach to uncover the genomic history of major adaptations in metazoan lineages. Trends Genet 23:533–9
Duan Z, Feller AJ, Toh HC, Makastorsis T, Seiden MV (1999) TRAG-3, a novel gene, isolated from a taxol-resistant ovarian carcinoma cell line. Gene 229:75–81
Elhaik E, Sabath N, Graur D (2006) The "inverse relationship between evolutionary rate and age of mammalian genes" is an artifact of increased genetic distance with rate of evolution and time of divergence. Mol Biol Evol 23:1–3
Flicek P et al (2007) Ensembl 2008. Nucleic Acids Res 36:0707–14
Frith MC et al (2006) The abundance of short proteins in the mammalian proteome. PLoS Genet 2:e52

Furney SJ, Alba MM, Lopez-Bigas N (2006) Differences in the evolutionary history of disease genes affected by dominant or recessive mutations. BMC Genomics 7:165

Gibbs RA et al (2007) Evolutionary and biomedical insights from the rhesus macaque genome. Science 316:222–234

Goldovsky L et al (2005) CoGenT + +: an extensive and extensible data environment for computational genomics. Bioinformatics 21:3806–3810

Harris MA et al (2004) The Gene Ontology (GO) database and informatics resource. Nucleic Acids Res 32:D258–261

Hurst LD, Smith NG (1999) Do essential genes evolve slowly? Curr Biol 9:747–750

Iwabe N, Kuma K, Miyata T (1996) Evolution of gene families and relationship with organismal evolution: rapid divergence of tissue-specific genes in the early evolution of chordates. Mol Biol Evol 13:483–493

Johnson ME et al (2001) Positive selection of a gene family during the emergence of humans and African apes. Nature 413:514–519

Kouprina N et al (2004) The SPANX gene family of cancer/testis-specific antigens: rapid evolution and amplification in African great apes and hominids. Proc Natl Acad Sci USA 101:3077–3082

Lander ES et al (2001) Initial sequencing and analysis of the human genome. Nature 409:860–921

Levine MT, Jones CD, Kern AD, Lindfors HA, Begun DJ (2006) Novel genes derived from non-coding DNA in Drosophila melanogaster are frequently X-linked and exhibit testis-biased expression. Proc Natl Acad Sci USA 103:9935–9939

Lynch M, Conery JS (2000) The evolutionary fate and consequences of duplicate genes. Science 290:1151–1155

Martinez-Morales JR, Henrich T, Ramialison M, Wittbrodt J (2007) New genes in the evolution of the neural crest differentiation program. Genome Biol 8:R36

Miyata T, Suga H (2001) Divergence pattern of animal gene families and relationship with the Cambrian explosion. BioEssays 23:1018–1027

Nurminsky DI, Nurminskaya MV, De Aguiar D, Hartl DL (1998) Selective sweep of a newly evolved sperm-specific gene in Drosophila. Nature 396:572–575

Ohno S (1984) Birth of a unique enzyme from an alternative reading frame of the preexisted, internally repetitious coding sequence. Proc Natl Acad Sci USA 81:2421–2425

Okamura K, Feuk L, Marques-Bonet T, Navarro A, Scherer SW (2006) Frequent appearance of novel protein-coding sequences by frameshift translation. Genomics 88:690–697

Pal C, Papp B, Lercher MJ (2006) An integrated view of protein evolution. Nat Rev Genet 7:337–348

Schittek B et al (2001) Dermcidin: a novel human antibiotic peptide secreted by sweat glands. Nat Immunol 2:1133–1137

Schmidt HA, Strimmer K, Vingron M, von Haeseler A (2002) TREE-PUZZLE: maximum likelihood phylogenetic analysis using quartets and parallel computing. Bioinformatics 18:502–504

Seo TK, Kishino H, Thorne JL (2004) Estimating absolute rates of synonymous and nonsynonymous nucleotide substitution in order to characterize natural selection and date species divergences. Mol Biol Evol 21:1201–1213

Stoye J, Evers D, Meyer F (1998) Rose: generating sequence families. Bioinformatics 14:157–163

Thompson JD, Higgins DG, Gibson TJ (1994) CLUSTAL W: improving the sensitivity of progressive multiple sequence alignment through sequence weighting, position-specific gap penalties and weight matrix choice. Nucleic Acids Res 22:4673–4680

Waterston RH et al (2002) Initial sequencing and comparative analysis of the mouse genome. Nature 420:520–562

Wootton JC, Federhen S (1996) Analysis of compositionally biased regions in sequence databases. Methods Enzymol 266:554–571

Yang Z (2007) PAML 4: phylogenetic analysis by maximum likelihood. Mol Biol Evol 24:1586–1591

Zhang G et al (2007) Identification and characterization of insect-specific proteins by genome data analysis. BMC Genomics 8:93

Chapter 4
Life-Cycle Features of Tumour Cells

Jekaterina Erenpreisa and Mark S. Cragg

Abstract *"The main rule for cancer cells is the absence of any rules" (Hanseman 1890).*

Although this statement was made over 100 years ago, it is still principally true today. However, in light of recent revelations concerning the nature and evolution of cancer and the explosion in stem cell research, it may be that we have been looking for the rules in the wrong place. Rather than looking for parallels in mitotically dividing somatic cells, here we discuss the notion that better rules may be found in the processes of germ-line cells. In this review we will revisit the centuries-old embryonal theory of cancer, explore a potential molecular rationale for its existence and discuss the hypothesis that the central characteristics of tumour cells are provided by their recapitulation of ancient protozoan life-cycle programmes which feature aspects of meiosis and sexual reproduction.

4.1 Introduction

Tumours undergo microevolution and progression driven by the twin motors of genetic variation and environmental selection. Unfortunately, the same pressures are often present after anticancer treatment, leading to the common scenario whereby the tumour mass initially regresses only to be followed by regrowth of (epi)genetically altered, resistant cells. Up until now, it was believed that tumour cells reproduce exclusively through traditional mitosis. The absence of clear rules

---

J. Erenpreisa
Latvian Biomedicine Research and Study Centre, Ratsupites str. 1, 1067 Riga, Latvia
katrina@biomed.lu.lv

M.S. Cragg
Tenovus Research Laboratory, Southampton University Hospital, Tremona Road, Southampton SO16 6YD, UK
msc@soton.ac.uk

for how mitosis was regulated in tumour cells was attributed to mutations in key players in the cell cycle checkpoints leading to the gradual loss of mitotic control. However, several surprising recent findings relating to the origins of cancer and regulation of cancer stem cells now lead us to question this view and consider whether the microevolution and propagation of tumours is in fact analogous to the macroevolution of species with features of life cycles. Amazingly, some of these notions were suggested over a century ago in the embryonal theory of cancer.

> *"Imagination is more important than knowledge"*
> *(A. Einstein)*

4.2 From Embryonal to Stem Cell Theories of Cancer

The embryological origins of cancer were suggested by several prominent scientists in the nineteenth century. Frequently this theory is connected with the name of Julius Cohnheim (1877–1880), who generalised the idea that tumours originate from embryonic cells retained in the adult. Since then the embryonal theory has resurfaced in various different forms, under the terms fertilisation, trophoblast, parthenogenetic and gametogenetic theories. These were united by the thought that carcinogenesis is intimately linked to sexual reproduction (developed and reviewed by Erenpreiss 1992, 1993). Although controversial, biochemical data exist to support this notion as numerous proteins normally only expressed in germ and embryonic cells are also observed in tumour cells (Knox 1976). More recent evidence for the embryonal theory of cancer shows that nuclei isolated from embryonal carcinoma and melanoma tumour cells can participate in the formation of embryonic stem cells when transferred into enucleated eggs and that these can prime the normal embryonal development of mice (Blelloch et al. 2004; Hochedlinger et al. 2004), confirming the previous experiments of this kind performed with the cells of frogs and mice (Illmensee and Mintz 1976). In turn, when embryonal tissues are introduced into syngenic adults, they convert directly into tumour (embryonal carcinoma) at the site of injection without any latent period (reviewed by Bradley 1990; Erenpreiss 1993). In fact, current tests for embryonic stem cells include the ability to form tumours (Cyranoski 2007). These striking data indicate two facets of tumour cells: (1) that tumour cells possess embryonal potential and (2) that ectopic embryonal tissue can lead directly to tumour formation.

Concerning the embryonal potential (foetality) of tumour cells it is intriguing to note that stem cells, the current favoured progenitor cell for tumourogenesis, possess many embryological features, including unlimited/extended proliferation capacity and totipotency or multipotency (Eckfeldt et al. 2005). Indeed, the latter is almost the definition of embryonality—the ability to create the whole organism or its tissues. Amazingly then, it seems that the centuries-old embryological theory of carcinogenesis is currently undergoing its latest renaissance as the stem cell theory of cancer.

4.3 Cancer Testes Antigens: Expression in Tumour and Germ Cells

Intriguingly, many tumours ectopically express the so-called cancer testes-associated antigens (CTA)—proteins expressed only in tumours and germ tissue such as testes, ovaries and placenta. At least some of these gene products belong to the gametogenetic and early embryogenesis developmental programmes (Old 2001, 2007; Simpson et al. 2005; Kalejs and Erenpreisa 2005; Jungbluth et al. 2007; Silva et al. 2007). Our own studies have shown that human lymphoma cells aberrantly express products usually only associated with germ cells. For example, we have observed ectopic expression of the meiotic kinase MOS, the meiotic cohesins Rec8 and STAG3, the meiotic recombinase DMC1 as well as the SCP1 and SCP3 proteins of the synaptonemal complex (Kalejs et al. 2006). Importantly, expression of these proteins is associated with mitotic catastrophe evoked by genotoxic stress and is correlated with the transient polyploidy induced in the tumour cells. Evidence of a potential link between CTA expression, embryonality and resistance to genotoxic stress was shown recently in embryonic stem cells (reviewed by Costa et al. 2007) with polyploid induction uncoupled from apoptosis. Furthermore, in embryonic stem cells, unlike in a more differentiated pathway, uncoupling of polyploidy from apoptosis is associated with expression of the germ-line gene *Oct*4 (Mantel et al 2007; Guo et al. 2008). Taken collectively, these molecular data strongly support the embryonality of tumour cells and provide clear links with polyploidy and the biology of stem cells. Next, we will consider the role of polyploidy in these phenomena.

4.4 From Mitosis to Polyploidy and Life Cycles

Polyploidy has a direct connection with macroevolution. Indeed, the majority of flowering plants and vertebrates have descended from polyploid ancestors. While the role of gene duplications in adaptive evolution is much discussed (reviewed by Otto 2007), the direct role of polyploidy in the evolution of the sexual process is less widely known. Conventional meiosis seen in higher organisms in fact evolved in protists, which represent a transitional evolutionary group linking prokaryotes with the multicellular eukaryotes (Whittaker 1969; Cavalier-Smith 2002). Cleveland (1947) suggested that meiosis evolved from mitosis in the polyploid forms of protozoans as a means of reducing chromosome number. Based upon his herculean work on 40 genera and 500 species of hypermastigote and polymastigote flagellates, Cleveland (1947) presented a scheme indicating progressive stages in the origin and direct evolution of meiosis from mitosis through the life cycles of these protozoans (Fig. 4.1). This was based upon his observations that certain haploid species occasionally produced polyploid forms and subsequently derived the mechanisms (initially asexual) to relieve the associated load of genetic instability through

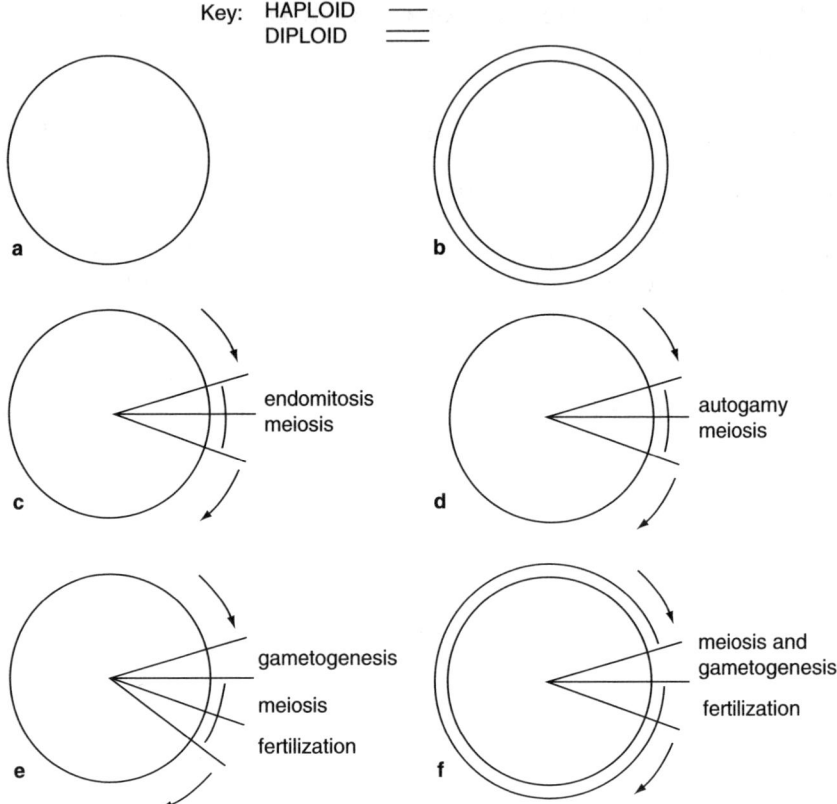

Fig. 4.1 The original scheme by Cleveland (1947) showing evolution of meiosis from mitosis: **a** haploidy in *Holomastigotoides*; **b** diploidy in *Holomastigotoides* (where bichromatid chromosomes go to the poles in each mitotic cycle); **c** endomitosis in *Barbulanympha*; **d** autogamy (fusion of nuclei) in *Barbulanympha* and *Saccinobaculus*; **e** production of gametes by cytoplasmic division and zygotic meiosis in *Trichonympha*; **f** gametic meiosis that precedes fertilisation in higher organisms

reduction mechanisms. These asexual ploidy cycles in protists were the forerunners of meiosis, and subsequently led to the evolution of the sexual process and sexual life cycles, as discussed by several prominent protistologists (Raikov 1995; Poljansky et al. 2000). Kondrashov (1994) has shown by mathematical analysis that asexual ploidy cycles reduce mutation load as compared with permanent diploidy or polyploidy and are thus likely to evolve in cases where it is advantageous to have more than one copy of the genome per cell. Returning to the scheme elaborated by Cleveland (Fig. 4.1), it is intriguing to note that in each subsequent evolutionary stage the events of the preceding stage are repeated, followed by an additional step. This allows us to identify the stage when meiosis-like processes developed and reveal that particular attention should be paid to endomitosis.

4.5 Endomitosis: The Earliest Evolutionary Analogue of Meiosis

Endomitosis was first described in polyploid cells of insects (Geitler 1937) and was originally defined as a cycle of chromosome condensation and partition without karyokinesis and dissolution of the nuclear envelope in polyploid cells. Cleveland suggested that endomitosis was perhaps the earliest evolutionary analogue of meiosis, appearing during the evolution of the polyploid protozoans. Similar observations in *Aulocatha*, including finding of the axial elements by electron microscopy (Grell and Ruthmann 1964), and discussions on its meaning between Grell and Cashon brought the last author to the conclusion that endomitosis is a cyclical process for pairing homologous chromosomes in the polyploid nucleus, in preparation for somatic reduction (reviewed by Raikov 1982). Interestingly, the relationship of endomitosis to meiosis has also been observed in certain snakes (Besak et al. 2003).

4.6 Endomitotic Tumour Cells Express Meiotic Kinases

Given its evolutionary origins as a forerunner to meiosis, it is therefore notable that endomitosis is a characteristic feature of mammalian tumour cells both in vitro and in vivo (Levan and Haushka 1953; Therman et al. 1986). Moreover, we found that the endomitotic nuclei of lymphoma cells contained the meiotic kinase MOS, which prevents degradation of cyclin B, and therefore may downregulate karyokinesis of these cells (Erenpreisa et al. 2005). In addition, we observed the ability of the polyploid cells to repair DNA double-strand breaks by homologous recombination (Ivanov et al. 2003) and the presence of meiotic cohesin Rec8 and meiotic recombinase DMC1 in the nuclei of polyploid cells induced by genotoxic stress (Kalejs et al. 2006). These observations suggest that these meiotic gene products may be associated with the participation of polyploid tumour cells in meiosis-like events.

4.7 Genetic Consequences of Reproductive Polyploidy in Tumour Microevolution

The genetic advantages and disadvantages of polyploidy in macroevolution are the subject of current debate (Comai 2005; Otto 2007). Clearly, genetic instability is a disadvantage of polyploidy but on the other hand if polyploidy per se was a dead end in macroevolution we would expect polyploid taxa to be at the ends of evolutionary strands and to be relatively species poor. Instead, several polyploidisation events are ancient and a number of them gave rise to species-rich groups (Nagl 1978). Therefore, it is likely that instable genomes and the extensive genomic repatterning which occurs in newly generated polyploids increases genetic variability, favouring adaptive evolution of descendent polyploid populations (Otto 2007).

Indeed, evolutionary studies have shown that as a result of polyploidy, new species can sometimes arise very rapidly, within only a few generations.

It seems that the same principles can be considered for the role of polyploidy in the microevolution of tumours. The available data indicate that although developmental endopolyploidy in mammals is rare and generative polyploidy is lethal (Otto 2007), this is not the case for mammalian tumours. In addition, the reproductive potential (clonogenicity) of endopolyploid tumour cells has been suggested from experimental work by several independent authors (reviewed by Rajaraman et al. 2005).

Polyploidy in tumours is often proclaimed as a source of increasing genetic instability and aneuploidy (Storchova and Pellman 2004; Nguyen and Ravid 2006). However, it should be noted that the presence of mechanisms balancing polyploid genomes has already been proposed (reviewed by Otto 2007). In addition, during the macroevolution of plants, polyploidy has in certain instances facilitated the shift from aneupolyploidy to diploidy through a 'triploid bridge' (Comai 2005). Clearly, mechanisms limiting aneusomy and counterbalancing increasing aneuploidy are necessary to prevent tumour progression leading to reproductive termination.

4.8 Do Tumour Cells Display Life-Cycle Behaviour Similar to that of Unicellular Protozoans?

One potential means of preventing loss of heterozygocity and accumulation of semideleterious mutations is through sexual reproduction or similar mechanisms where genetic exchange between homologous chromosomes is possible. Here, meiosis-like recombination between homologues is preferred over mitotic recombination between sister chromatids. Meiotic cohesins co-operating with monopolin provide this function during meiosis (Nasmyth 2001; Lee and Orr-Weaver 2001) and therefore the expression of the meiotic cohesin Rec8 and meiotic recombinase DMC1 in polyploid tumour cells provides an important hint of possible meiosis-like recombination in them.

Following genotoxic insult, p53-deficient tumour cells undergo a chain of events involving several kinds of aberrant cell divisions. Some of these divisions are interspersed with autologous cell or nuclei fusions (Chu et al. 2004; Erenpreisa et al., unpublished results). These divisions and fusions which occur over a period of approximately 2 weeks are cyclic and therefore seem highly reminiscent of a life cycle. Although much less frequent, the same process is observed in untreated tumour cells (unpublished results). Our hypothesis is that 'mitotic catastrophe' which is present at very low levels in untreated tumour cells, and elevated after genotoxic stress, serves as a bridge between mitosis and polyploidy, providing the mechanism through which tumours enter this life-cycle-like programme (Erenpreisa and Cragg 2007). Intriguingly, many of the features of this process are reminiscent of protozoan life cycles.

Protists are mostly unicellular organisms which also undergo life cycles and represent a giant evolutionary laboratory. They provide variations of all of the known basic life-cycle processes—from mitosis, meiosis and syngamy to haploidy, diploidy and amphyploidy, including cycling polyploidy (Raikov 1982; Ivanov and Kolchinsky, 2000). Amazingly, many of the hallmark protozoan life-cycle features have also been observed in mammalian tumour cells undergoing polyploidy. These are (1) the appearance of cilia in dividing human polyploid lymphoma cells (Fig. 4.2a, b), (2) the most primitive type of mitosis—pleuromitosis—whereby the nuclear envelope is preserved (Fig. 4.2f), (3) chromatin extrusion (diminution) from viable polyploid cells (Erenpreisa et al. 2000), (4) segregation of cytoplasm into endoplasm and ectoplasm and ameboidisation (Fig. 4.2d), (5) polycellurisation with the formation of karyoplasts (Roumier et al. 2005; Erenpreisa et al., unpublished results) and (6) cycling polyploidy (Sundaram et al. 2004; reviewed by Erenpreisa and Cragg 2007).

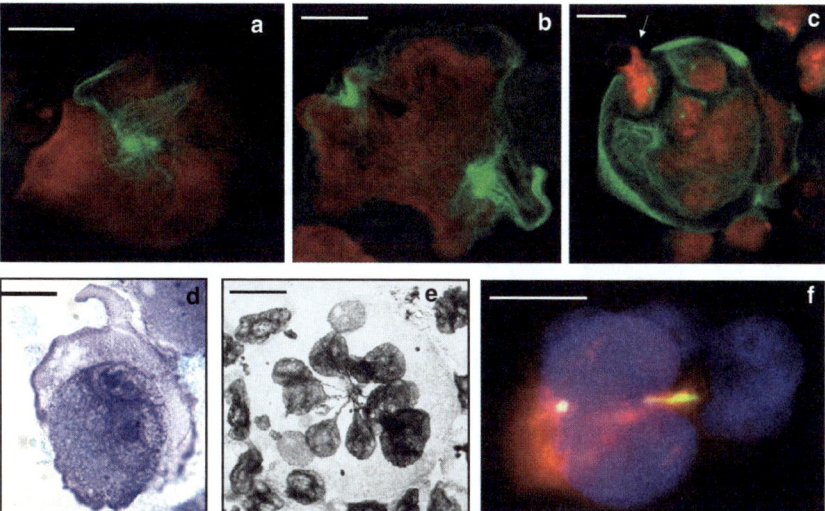

Fig. 4.2 Endopolyploid human tumour cells displaying features of the protozoan life cycle. **a, b** The appearance of flagella during the transition from a monopolar to a bipolar spindle prior to depolyploidisation (Namalwa lymphoma cell line). **c** Segregation of a polyploid giant cell into subcells (Namalwa). **d** A binuclear polyploid HeLa carcinoma cell showing segregation of its cytoplasm. It demonstrates the outer layer with a pseudopodal projection and the inner layer, rich with ribosomal material. A split is seen between the endoplasm and the ectoplasm; pale ectoplasm is often seen blebbing from these cells and is later discarded. **e** A multisegmented polyploid HeLa nucleus with radial nuclear bridges. **f** Bipolar spindle and apparent segregation of a polyploid nucleus (or nuclei) with preserved nuclear envelope. Vivid condensation of chromosomes is not apparent (Namalwa). This type of division is reminiscent of the most primitive type of mitosis—pleuromitosis. All cells were irradiated with a single 10-Gy dose of irradiation and fixed after the given time: **a, b** after 7 days; **c** 13 days; **d** 11 days; **e** 9 days; **f** 7 days. **a–c** Stained for β-tubulin (fluorescein isothiocyanate, *green*) and propidium iodide (*red*). **d, e** Toluidine blue pH 5.0, after partial RNA extraction. **f** Stained for γ-tubulin (fluorescein isothiocyanate, *green*) plus α, β-tubulin (tetramethylrhodamine isothiocyanate, *red*) and 4′,6-diamidino-2-phenylindole (*blue*). Bars 10 µm

4.9 Role of p53

If we accept that tumour cells are following a life-cycle programme akin to that observed in protozoans, the obvious question becomes how are these ancient pathways accessed? As detailed earlier, the how may be through 'mitotic catastrophe' and endomitosis at least in part, but following on from this we must question why these processes themselves are prominent in tumours. The answer to this may be found in the following rationale. Tumour cells display a large number of different genetic and epigenetic mutations enhancing their proliferation and survival. However, they have one change in common—the inactivation of p53, the guardian of the genome. The p53 tumour suppressor gene appeared during evolution from the p63 family following the divergence of teleosts and amphibians (Saccone et al. 2002). The function of this gene is disabled during tumour progression in a variety of different ways in practically all tumours (Kastan 2007). The importance of this inactivation may be twofold. First, in normal cells, p53 upregulation after genotoxic damage results in apoptosis, preventing access to mitotic catastrophe, polyploidy and subsequently to endomitosis. Second, and potentially more relevant for tumorogenesis and progression, p53 binds to the promoter of the master self-renewal stem cell gene *Nanog*, inhibiting its expression and inducing differentiation (Lin et al. 2005). Overexpression of p53 enhances apoptosis, differentiation and cell senescence (Dimri 2005). Therefore, inhibition of p53 function may shift the tumour stem cell towards survival, self-renewal and dedifferentiation. Thus, in the absence of tumour suppression by p53, stemness is favoured and access to mitotic catastrophe and endomitosis is provided.

4.10 Adaptive Paleogenesis in Tumours

In embryonal stem cells the chromatin conformation is open (Eckfeldt et al. 2005) and 75–80% of genes are transcriptionally initiated (Guenther et al. 2007). Adult stem cells can acquire the embryonal-stem-cell-like transcription profile under various experimental or stress conditions (Tam et al. 2007; Cyranoski 2007). Why should this matter? According to the biogenetic law of Haeckel (1866), ontogenesis is a recapitulation of phylogenesis. Although an oversimplification, and later modified by Severtsov (1939) and others to incorporate the idea that paleogenetic features in ontogenesis serve the adaptive evolution of new species, today it is recognised that "on a fundamental level, Haeckel was correct" (Richardson et al. 1998). Taking this notion a step further, we hypothesise that the transcription priming and open chromatin conformation which occur in tumour stem cells opens the genetic memory for recapitulation of paleogenetic elements of the life cycles of protozoan ancestors. Similar to the role of phylogenetic embryogenesis in the macroevolution of species, this recapitulation would represent an adaptive response in tumour microevolution assigning selected tumour clones with many potential survival advantages.

4.11 Conclusion

It seems that after over a century of scientific endeavour the theory of carcinogenesis has come full circle, once again recognising embryonality (stemness) as the most essential biological feature of the tumour cell. Thus, investigations are beginning to move away from studies of the aberrant regulation of mitosis in tumour cells to the more complex rules of life cycles. This shift allows us to explore analogies between the microevolution of tumours and macroevolution of species. Polyploidy, it seems, represents a very important driving force for both.

Acknowledgements We are grateful to A.P. Anisimov (Vladivostok), who drew our attention to the segregation of cytoplasm in HeLa cells, and to Bodo Liebe (Berlin) and Martins Kalejs (Riga), who prepared specimens for Fig. 2f. Some confocal pictures were processed with the excellent assistance of Roger Alston (Southampton). We are very grateful to Helmut Zacharias (Langwendel) for discussion of some aspects of the manuscript.

References

Besak ML, Besak W. Pereira A (2003) Somatic pairing, endomitosis and chromosome aberrations in snakes (Viperidae and Colubridae). Ann Braz Acad Sci 75: 285–300

Blelloch RH, Hochedlinger K, Yamada Y, Brennan C, Kim M, Mintz B, Chin L, Jaenisch R (2004) Nuclear cloning of embryonal carcinoma cells. Proc Natl Acad Sci USA 101:13985–13990

Bradley A (1990) Embryonic stem cells: proliferation and differentiation. Curr Opin Cell Biol 2:1013–1017

Cavalier-Smith T (2002) The phagotrophic origin of eukaryotes and phylogenetic classification of Protozoa. Int J System Evol Microbiol 52: 297–354

Chu K, Teele N, Dewey MW, Albright N, Dewey NW (2004) Computerized video time lapse study of cell cycle delay and arrest, mitotic catastrophe, apoptosis and clonogenic survival in irradiated14-3-3s and CDKN1A (p21) knockout cell lines. Radiat Res 162: 270–286

Cleveland LR (1947) The origin and evolution of meiosis. Science 105:287–289

Cohnheim J (1877–1880) Vorlesungen über allgemeine Pathologie. Ein Handbuch fur Ärzte und Studierende. vols 1–2, Hirschwald, Berlin

Comai L (2005) The advantages and disadvantages of being polyploidy. Nat Rev Genet 6:836–846

Costa FF, Le Blanc K, Brodin B (2007) Concise review: cancer/testis antigens, stem cells, and cancer. Stem Cells 25:707–711

Cyranoski D (2007) Race to mimic human embryonic stem cells. Nature. doi:10.1038/450462a

Dimri GP (2005) What has senescence got to do with cancer? Cancer Cell 7:505–512

Eckfeldt CE, Mendenhall EM. Verfaillie CM (2005) The molecular repertoire of the 'almighty' stem cells. Nat Rev Mol Cell Biol 6:626–637

Erenpreisa J, Cragg MS (2007) Cancer: a matter of life cycle? Cell Biol Int 31:1507–1510

Erenpreisa J, Kalejs M, Cragg MS (2005) Mitotic catastrophe and endomitosis in tumour cells: an evolutionary key to a molecular solution. Cell Biol Int 29:1012–1018

Erenpreisa JA, Cragg MS, Fringes B, Sharakhov I, Illidge TM (2000) Release of mitotic descendants by giant cells from irradiated Burkitt's lymphoma cell line. Cell Biol Int 24:635–648

Erenpreiss J (1992) Gametogenesis as a molecular model for cancerogenesis. A current view of the embryological theory of cancer. Proc Latv Acad Sci Ser B 3:55–63

Erenpreiss J (1993) Current concepts of malignant growth. Zinatne, Riga

Geitler L (1937) Die Analyse des Kernbaus und der Kernteilung der Wasserlaufer Gerris lateralis un Gerris lacustris (Hemiptera Heroptera) un die Somadifferenzierung. Z Zellforsch 26:641–72

Grell KG, Ruthmann A (1964) On the karyology of the radiolarian Aulocantha Scolymantha and fine structure of its chromosomes. Chromosoma 15:185–211

Guenther MG, Levine SS, Boyer LA, Jaenisch R, Young RA (2007) A chromatin landmark and transcription initiation at most promoters in human cells. Cell 130:77–88

Guo Y, Mantel C, Hromas RA, Broxmeyer HE (2008) Oct 4 is critical for survival/antiapoptosis of murine embryonic stem cells subjected to stress. Effects associated with STAT3/Survivin. Stem Cells 26:30–34

Haeckel E (1866) Generelle Morphologie der Organismen, vol 2. Allgemeine Entwickelungsgeschichte der Organismen. Reimer, Berlin

Hansemann D (1890) Uber asymmetrische Zelltheilung in Epithelkrebsen und deren biologische Bedeutung. Arch Pathol Anat Physiol Klin Med 119:299–326

Hochedlinger K, Blelloch R, Brennan C, Yamada Y, Kim M, Chin L, Jaenisch R (2004) Reprogramming of a melanoma genome by nuclear transplantation. Genes Dev 18:1875–1885

Illidge T, Cragg M, Fringe B, Olive P, Erenpreisa J (2000) Polyploid giant cells provide a survival mechanism for p53 mutant cells after DNA damage. Cell Biol Int 24:621–633

Illmensee K, Mintz B (1976) Totipotency and normal differentiation of single teratocarcinoma cells cloned by injection into blastocysts. Proc Natl Acad Sci USA 73:549–553

Ivanov A, Cragg MS, Erenpreisa J, Emzinsh D, Lukman H, Illidge TM (2003) Endopolyploid cells produced after severe genotoxic damage have the potential to repair DNA double strand breaks. J Cell Sci 116:4095–4108

Ivanov AV. Kolchinsky EI (2000) Pathways and principles of the evolution. In: Alimov AF (ed) Protista, part I. Nauka, St Petersburg, pp 29–85

Jungbluth AA, Silva WA Ir, Iversen K, Frosina D, Zaidi B, Coplan K, Eastlake-Wade SK, Castelli SB, Spagnoli GC, Old LJ, Vogel M (2007) Expression of cancer-testis (CT) antigens in placenta. Cancer Immun 7:15

Kalejs M, Erenpreisa J (2005) Cancer/testis antigens and gametogenesis: a review and "brainstorming" session. Cancer Cell Int 5:4

Kalejs M, Ivanov A, Plakhins G, Cragg MS, Emzins Dz , Illidge TM, Erenpreisa Je (2006) Upregulation of meiosis-specific genes in lymphoma cell lines following genotoxic insult and the induction of mitotic catastrophe. BMC Cancer 6:6

Kastan MB (2007) Wild-type p53: tumours can't stand it. Cell 128:837–840

Knox WE (1976) Enzyme patterns in fetal, adult and neoplastic rat tissues. 2nd edn. Karger, Basel, p 359

Kondrashov AS (1994) The asexual ploidy cycle and the origin of sex. Nature 370:213–216

Lee JY, Orr-Weaver TL (2001) The molecular basis of sister chromatid cohesion. Annu Rev Cell Dev Biol 17:753–777

Levan A, Hauschka TS (1953) Endomitotic reduplication mechanisms in ascites tumors of mouse. J Natl Cancer Inst 14:1–43

Lin T, Chao C, Saito S, Mazur SJ, Murphy ME, Appella E, Yaung X (2005) p53 induces differentiation of mouse embryonic stem cells by suppressing Nanog expression. Nat Cell Biol 7:165–171

Mantel C, Guo Y, Lee MR, Kim M-K, Han M-K. Shibayama H., Fukuda S, Yoder MC, Pelus LM (2007) Checkpoint-apoptosis uncoupling in human and mouse embryonic stem cells: a source of karyotpic instability. Blood 109:4518–4527

Nagl W (1978) Endopolyploidy and polyteny in differentiation and evolution. North Holland, Amsterdam

Nasmyth K (2001) Disseminating the genome: Joining, resolving, and separating sister chromatids during mitosis and meiosis. Annu Rev Genet 35:673–745

Nguyen HG, Ravid K (2006) Tetraploidy/aneuploidy and stem cells in cancer promotion: the role of chromosome passenger proteins. J Cell Physiol 208:12–22

Old LJ (2001) Cancer/testis (CT) antigens—a new link between gametogenesis and cancer. Cancer Immun 1:1

Old LJ (2007) Cancer is a somatic cell pregnancy. Cancer Immun 7:19
Otto SP (2007) The evolutionary consequences of polyploidy. Leading edge review. Cell 131:452–462
Poljansky GI, Sukhanova KM, Karpov SA (2000) General characteristics of protists. In: Alimov AF (ed) Protista, part I. Nauka, St Petersburg, pp 145–184
Raikov IB (1982) The protozoan nucleus, morphology and evolution. Springer, Vienna
Raikov IB (1995) Meiosis in protists: recent advances and persisting problems. Eur J Protistol 31:1–7
Rajaraman R, Rajaraman MM, Rajaraman SR, Guernsey DL (2005) Neosis—a paradigm of self-renewal in cancer. Cell Biol Int 29:1084–1097
Richardson MK, Hanken J, Selwood L, Wright GM, Richards RJ, Pieau C, Raynaud A (1998) Haeckel, embryos, and evolution. Science 280:983, 985–986
Roumier T, Valent A, Perfettini JL, Metivier D, Castedo M, Croemer G (2005) A cellular machine generating apoptosis-prone cells. Cell Death Differ 12:91–93
Saccone C, Barome PO, D'Erchia AM, D'Errico I, Pesole G, Sbisa E' Tullo A (2002) Molecular strategies in metazoan genomic evolution. Gene 300:195–201
Severtsov AN (1939) Morphological regularities of evolution. Academy of Sciences USSR, Moscow, p 538
Silva WA Jr, Ritter GS, Cohen CR, Hsu M, Jungbluth AA, Altorki NK, Chen YT, Old LJ, Simpson AJ, Caballero OL (2007) PLAC1, a trophoblast-specific cell surface protein, is expressed in a range of human tumors and elicits spontaneous antibody responses. Cancer Immun 7:18
Simpson AJ, Caballero OL, Jungbluth A, Chen YT, Old LJ (2005) Cancer/testis antigens, gametogenesis and cancer. Nat Rev Cancer 5:615–625
Storchova Z, Pellman D (2004) From polyploidy to aneuploidy, genome instability and cancer. Nat Rev Mol Cell Biol 5:45–54
Sundaram M, Guernsey DL, Rajaraman MM, Rajaraman R (2004). Neosis: a novel type of cell division in cancer. Cancer Biol Ther 3:207–218
Tam WL, Ang Y-S, Lim B (2007) The molecular basis of ageing in stem cells. Mech Ageing Dev 128:137–148
Therman E, Sarto G, Kuhn EM (1986) The course of endomitosis in human cells. Cancer Genet Cytogenet 19:301–310
Whittaker RH (1969) New concepts of kingdoms of organisms. Science 163:150–160

Chapter 5
General Evolutionary Regularities of Organic and Social Life

Valeria I. Mikhalevich

Abstract The main evolutionary regularities were firstly and more fully studied in biological sciences. An important contribution to the subject was made by the Russian evolutionary school, elaborating the evolutionary ideas mostly on multicellular organisms. Protista (mainly Infusoria) in these investigations were only slightly touched upon. The present study represents the results of a comparative morphological analysis of the vast protistan group Foraminifera, showing the manifestation of the evolutionary regularities at the unicellular level. Research concentrated on the major regularities—polymerization, differentiation, and integration—representing the mainstream of evolutionary development and permitting the structures to achieve a new higher level of organization. The phenomenon of aromorphoses is partly touched upon. It is also shown that these processes have a general character and can be applied to the social level of life as well.

5.1 Introduction

The investigation of the regularities of evolutionary development began with the classic works of Cuvier (1801), Lamark (1804), and Darwin (1859), and a significant contribution to the study of evolutionary mechanisms was made by the scientists of the Russian school (Berg 1922; Severtzov 1925, 1939; Schmalchgausen 1939, 1946; Dogiel 1929, 1954; Beklemishev 1964; Golubowski 1994; and others) and multiple recent studies. All of them were based mainly on biological objects and predominantly on multicellular organisms.

V.I. Mikhalevich
Zoological Institute, Russian Academy of Sciences, Universitetskaya nab. 1,
St. Petersburg 199134, Russia
mikha@JS1238.spb.edu

The investigation of the evolutionary regularities of unicellular organisms was difficult because of the minute sizes of these objects and the scarcely lack of structural morphological characters.

Foraminifera—the vast group of protists—represents in this regard one of a lucky exceptions owing to the variability and complexity of their skeletons, which is unusual at the unicellular level. The abundant occurrence of their agglutinated and calcareous shells in the geological strata since the Cambrian also provides an opportunity to study their evolutionary development through time. Foraminifera are sea animals that have existed to the present and whose soft cell is enclosed by an agglutinated or calcareous (rarely tectinous) shell. The animals communicate with the environment through an opening in the shell—their aperture—protruding and spreading the cytoplasmic reticulopodia (Figs. 5.1 and 5.2). The cytoplasm of the living animals owes is color to the different symbiontic algae.

In Fig. 5.2 there are shown the main phyletic lines of this group: classes Astrorhizata, Spirillinata, Nodosariata, Miliolata, and Rotaliata (according to the classification proposed by Mikhalevich 1980, 1981, 1992, 1999, 2000, 2004 2005). It was supposed that their early representatives lacking a hard test existed much earlier (Mikhalevich 1981, 2000; Mikhalevich and Debenay 2001). Molecular data of Pawlowski et al. (2003) have clocked that the early foraminiferal evolution occurred between 690 and 1,150 million years ago. These data are in agreement with the data of other authors (Hengeveld and Fedonkin 2004) on the origin of the early eukaryots in the Neoproterozoic. Nevertheless their skeletons, as well as the skeletons of the other animal groups, have been well preserved since the Cambrian. Many of the representatives of each of the classes shown in Fig. 5.2 are still living in the seas. Class Rotaliata has the most recent origin (Mesozoic, though its earlier

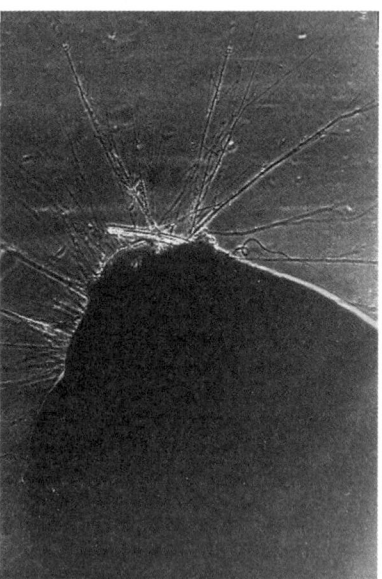

Fig. 5.1 Apertural part of the living *Massilina secans* (d'Orbigny), 1826, littoral Ile d'Yeu, Biscay Bay, 1998, with extruded reticulopodia (×50)

5 General Evolutionary Regularities of Organic and Social Life

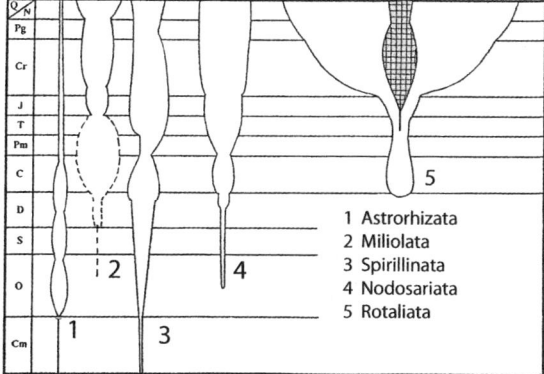

Fig. 5.2 The main phyletic foraminiferal lines (classes) according to the classification of Mikhalevich (1992) (scheme, not to scale) (the *dotted line* outlines the previous heterogenous taxon Textulariina of Loeblich and Trappan 1987). Note the unilocular shells at the beginning of Nodosariata and Miliolata lines, pseudo-two-chambered shells with the long tubular second chamber at the beginning of Spirillinata and Miliolata lines, and supermultichambered shells at the end of Miliolata and Rotaliata lines

more primitive representatives are known from the Carbonaceous), Nodosariata and Miliolata are known from the Ordovic, and Spirillinata are known from the Cambrian. Astrorhizata, which are the more primitive and always unilocular forms, existed in the Pre-Cambrian (Fig. 5.3).

After the first study of d'Orbigny, in 1826, foraminiferal classification was based on the features of the skeletons of Foraminifera. From Schultze (1854) to the most widely used classification of Loeblich and Tappan (1964, 1987), predominant significance was given to the peculiarities of the composition of the shell wall (agglutinated with sand particles attached or calcareously secreted) and its structure. The possibilities presented by the introduction of electronic microscopy permitting ultrastructural studies strongly strengthened such an approach. As a result, many of the previous foraminiferal taxonomic groups were divided into a number of different ones which were placed in the remote phyletic lines in spite of their nearly full shell structure isomorphism. The isomorphic forms were regarded as the consequence of the convergent evolutionary development. Thus, all the agglutinated forms, including all the existing morphotypes of the foraminiferal shell, were united in one taxon—Textulariina (Fig. 5.2). Spirillinata and Rotaliata lines were split into several groups of equally high taxonomic rank.

After the thorough comparative morphological analysis of all the main foraminiferal taxa, a new classification scheme (Fig. 5.2) was proposed (Mikhalevich 1992, 1998, 1999, 2000), based mainly on the morphological features of the foraminiferal shell regarding the composition and ultrastructure of the shell wall as having important but subordinate meaning. Under such an approach the agglutinated and calcareous isomorphs showing profound similarity in their shell and apertural structures and similar tendencies in their onthogenetic and phylogenetic development were placed within one phyletic line (class), their resemblance being

Fig. 5.3 The development of different foraminiferal classes in geological history (in Miliolata, Fusulinoida are included, *dotted line*; in Rotaliata, planktonic Globigerinana are shown *checked*). (Modified after Mikhalevich 2000)

considered a result of the close relationship rather than of convergence (Mikhalevich 2000, 2004). The agglutinated forms may represent the earlier phylogenetic stages of the development of their calcareous isomorphs or their sister groups. Some of the new data of the foraminiferal cell, especially those of the character of the nuclear apparatus, also support the separating of these five classes. These cytological characters were used as the taxonomic features of the classes for the first time. Further discussion in this chapter is based on this new classification scheme which in its four lines (Astrorhizata, Miliolata, Spirillinata, Rotaliata) was later supported by the molecular data of the Pawlowski school (Pawlowski et al. 2003) (data on Nodosariata are still absent).

Each of the five classes could be specified in terms of its special morphological features, such as plan of the structure of the shell connected with the functioning of its parts, predominant modes of coiling, characteristic form of their chambers, position of the aperture and its outer and especially inner apertural structures, and development of additional systems (additional apertures, different types of integrative systems).

5.2 Processes of Polymerization in Foraminiferal Development

The main evolutionary processes having some specific features in these different classes (Fig. 5.2) occurred independently and in parallel in each of them during the geological time. Thus, the majority of the classes began their development from the unilocular (Astrorhizata, Nodosariata, Miliolata) or pseudo-two-chambered (subspherical proloculus followed by a long tubular second chamber—Miliolata, Spirillinata) forms (Fig. 5.2). Astrorhizata, including forms with only an agglutinated or tectinous shell wall (subclass Lagynana), did not get beyond the unilocular level of organization, being unilocular throughout. In the rest of the classes evolutionary development resulted in the creation of multichambered shells at the ends of their evolutionary branches (Spirillinata, Nodosariata, and Miliolata) or from the very beginning of their development (Rotaliata) (Fig. 5.2). In the more primitive multichambered forms there are three to seven chambers (Fig. 5.4, group A, nos. 3, 4); more usually there are 25–30 (Fig. 5.4, group A, nos. 6–15). In the two phyletic lines–in Rotaliata and Miliolata (Fig. 5.2), supermultichambered shells are known, where there may be up to thousands of chambers (Fig. 5.4, group A, nos. 16, 21–23).

What were the main ways and the main regularities of foraminiferal development that resulted in such variable and complex forms? The first stage of foraminiferal evolution manifested in the transition from the unilocular to the multichambered and even supermultichambered shell was the process of polymerization (the increase in similar homologous formations in the organism). Polymerization could take place in all living organisms—for instance, in the Metazoa cells are polymerized. The peculiarity of the Foraminifera as well as other Protista is that this process occurs within a single cell. In such tiny animals, pure in their structural features, polymerization is the basic primary process permitting further development—the second stage of the evolutionary process—the process of differentiation, and providing the base for it.

Some other skeletal structures in Foraminifera were the consequence of the processes of polymerization. Thus, the apertural openings of the main aperture of the shell could also be polymerized in the different phyletic lines of Foraminifera (Fig. 5.2, group B, nos. 1–9). And above the main aperture in some advanced foraminiferal genera additional (supplementary) apertures appeared (Fig. 5.4, group C, nos. 1–8). Depending on their disposition in the shell they are called sutural, umbilical, or peripheral supplementary apertures. Supplementary apertures are widely represented in the class Spirillinata and especially in Rotaliata; in Miliolata and Nodosariata only one or two genera having supplementary apertures are known. Supplementary apertures represent an example of multiplication not only of the apertural openings but also of the number of apertural systems and at the same time

78   V.I. Mikhalevich

of the differentiation of the apertural system into the main and supplementary ones. The multiplication of all types of apertural openings provides better communication of the organism with the environment.

Some cytoplasmic structures in Foraminifera were also the consequence of the processes of polymerization in their different phyletic lines (Fig. 5.4, group D, nos. 1–6). In the more primitive unilocular agglutinated and tectinous shells of the subclasses Astrorhizana and Lagynana (class Astrorhizata), the cell may have one or several (polymerized) nuclei (Fig. 5.4, group D, nos. 1, 2). In the representatives of

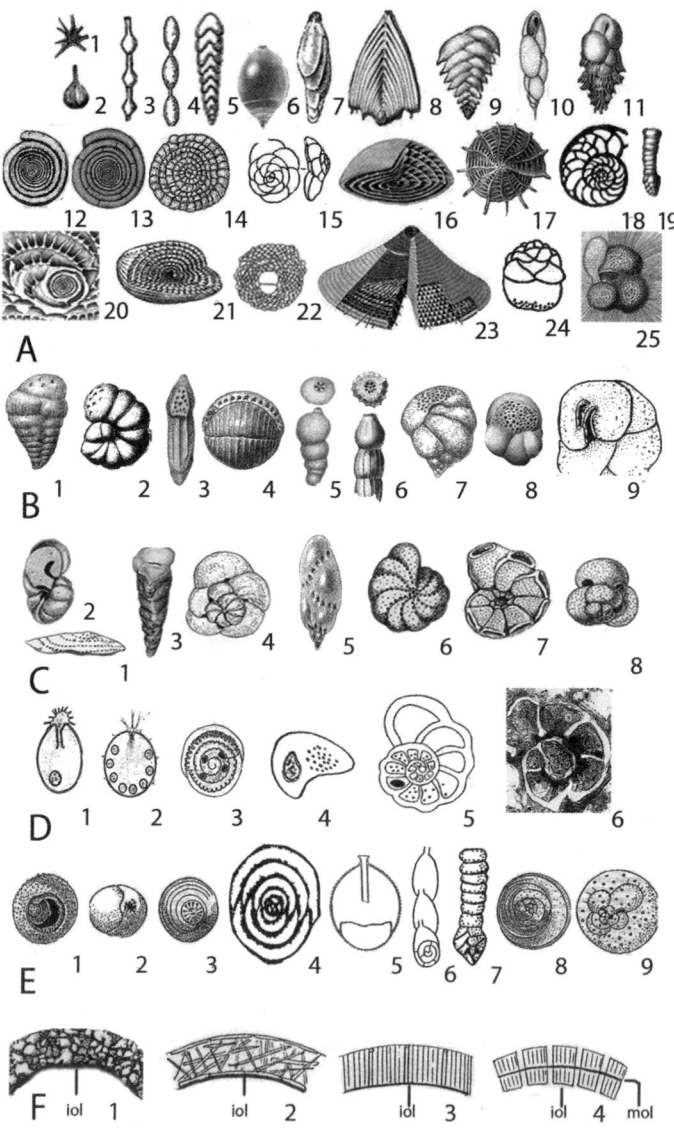

5 General Evolutionary Regularities of Organic and Social Life

the classes Spirillinata, Miliolata, and Rotaliata having multichambered calcareous shells the nuclei are always polymerized (Fig. 5.4, group D, nos. 3–5). The nuclear apparatus of Nodosariata has not been investigated yet. Polyploidization studied in Foraminifera only in nuclei of some representatives of the class Rotaliata (Zech 1964; Voronova and Mikhalevich 1985) represents an example of the polymerization of genomes. This is also the case in the nuclei of some other protistan groups (e.g., Radiolaria) whose genomes are also polymerized.

5.3 Processes of Differentiation in Foraminiferal Development

The process of differentiation becomes possible when there are already preformed structures consisting of numerous elemental units (Fig. 5.4, group A). The elemental units of the foraminiferal shell are represented by their chambers. The

---

Fig. 5.4 The details of the structure of the foraminiferal shells, apertures, wall ultrastructure, and nuclear apparatus (not to scale). *A* polymerization and differentiation of the chambers of the shell: 1, 2 unilocular shells—*Astrorhiza* (1), *Lagena* (2); 3, 4 polymerized undifferentiated chambers of equal size—*Grigelis* (3), *Saccamminopsis* (4); 5-25 chambers differentiated in size—*Lunucammina* (5), *Pseudonodosaria* (6), *Gorisella* (7), *Frondicularia* (8), *Bolivinella* (9), *Stainforthia* (10), *Bulimina* (11), *Cornuspira* (12; circular proloculus followed by tubular chamber), *Dolosella* (13), *Loeblichia* (14), *Discorbis* (15), *Nummulites* (16), *Elphidium* (17), *Heterostegina* (18), *Clavulina* (19), *Discospirina* (20), *Neoschwagerina* (21), *Lepidocyclina* (22), *Orbitolina* (23), *Tretomphalus* (24), *Globigerinoides* (25); 18-25 the chambers are differentiated also in form; 21-25 the chambers are differentiated also in function; 21-23 initial large embryonal chamber; 24 a big floating chamber at the lower part of the shell; 25 the last elongated brood chamber (*light in color*). *B* polymerization of the main apertures: *Cribrostomum* (1), *Haplophragmella* (2), *Hauerina* (3), *Borelis* (4), *Marginulina* (5), *Amphimorphina* (6), *Sporobuliminella* (7), *Neocribrella* (8), *Anticleina* (9). *C* additional apertures (as *black openings*): *Trocholinopsis* (1), *Polystomammina* (2), *Norvanganina* (3), *Toretammina* (4), *Virgulinella* (5), *Cribroelphidium* (6), *Almaena* (7), *Globigerinoides* (8); 1 Spirillinata; 2-8 Rotaliata; (1, 4, 5, 6, 8 sutural apertures; 2 umbilical apertures; 3, 7 peripheral apertures. *D* polymerization and differentiation of the nuclei: *Iridia* (1), *Myxotheca* (2), *Patellina* (3), *Quinqueloculina* (4), *Cibicides* (5, 6); 1, 2 Astrorhizata; 3 Spirillinata; 4 Miliolata; 5, 6 Rotaliata; 1 single nucleus; 2, 3 polymerized nuclei of equal size; 5 polymerized nuclei differentiated into the somatic macronucleus (*large, black*) and generative micronuclei (multiple); 6 disintegrating macronucleus passing through the foramens (*black*). *E* oligomerization of the previously polymerized chambers: *Ammosphaerulina* (1), *Idalina* (2,3), *Paleopatellina* (4), *Bombulina* (5), *Dimorphina* (6), *Clavulina* (7), *Neoconorbina* (8), *Orbulina* (9); 1-3 Miliolata; 4 Spirillinata; 5 Nodosariata; 6-9 Rotaliata; 1, 2, 9 last chamber enveloping the previous multichambered shell; 4, 5 diminished number of chambers in the final uniserial part; 4, 8 diminished number of chambers in the last coils of the trochospirally coiled shells). *F* the ultrastructure of the calcareous shell wall: microgranular wall with the randomly situated grains of variable size occurring in the lower representatives of all the foraminiferal classes (1), porcellaneous wall with randomly oriented needle crystals characteristic for Miliolata (2), hyaline wall with crystals situated perpendicularly to the shell surface characteristic for Nodosariata (3), hyaline wall with perpendicularly organized crystals situated in two layers and separated by the additional organic layer characteristic for the Rotaliata (4). *iol* inner organic lining serving as a wall matrix, *mol* middle organic lining between two lamellae of the Rotaliata wall. (*D* 1-5 after Mikhalevich 2005, 6 after Mikhalevich 2000; *E* after Mikhalevich 2000, modified)

chambers are separated from each other by the septa. Both chambers and septa create the compartmentalization of the shell space and of the cytoplasm in these chambers. In the simplest case the polymerized chambers are similar, equal in size and form (Fig. 5.4, group A, nos. 3, 4). But the multiple structures provide further opportunities—these multiple chambers could be differentiated. The process of differentiation is the second step in the evolutionary development. Thus, in Metazoa cells are differentiated into different tissues. In Foraminifera chambers may be differentiated in size (the simplest case) (Fig. 5.4, group A, nos. 5–25) and also in form (Fig. 5.4, group A, nos. 18–25). In some advanced genera the chambers of Foraminifera are differentiated in function (Fig. 5.4, group A, nos. 21–23, embryonal chambers; no. 24, floating chamber; no. 25 brood chamber). In this latter case the process of differentiation is tightly bound with the process of specialization. This process also took place in parallel in different classes and at different geological times. Embryonic chambers were formed in the higher representatives of the classes Spirillinata (Cretaceous orbitolins), Miliolata (Paleozoic fusulinids, Cretaceous alveolinids), and Rotaliata (Paleocene nummulitids) (Fig. 5.4, group A, nos. 21–23). Differentiation of the chamber form at the different stages of ontogenetic development occurred in each of the classes possessing multichambered forms. *Heterostegina* among Rotaliata, *Discospirina* among Miliolata, and *Orbitolina* among Spirillinata (subclass Ammodiscana) could be named as the best examples of such differentiation (Fig. 5.4, group A, nos. 18, 20, 23). Such functionally specialized chambers as the floating and brood chambers are known only within the more advanced foraminiferal class Rotaliata (Fig. 5.4, group A, nos. 24, 25).

The process of differentiation is often combined with the process of oligomerization when similar multiple structures become diminished in number. Thus, in each of the four classes with multichambered foraminiferal shells in many advanced genera the number of chambers in their last whorls in the final growth stages is less than that in their earlier whorls (Fig. 5.4, group E, nos. 1–9); for example, one chamber compared with five to ten chambers in their initial whorls. This is demonstrated in the shells with final subcircular (Fig. 5.4, group E, no. 8) or annular chambers or in the shells forming the uniserial part (Fig. 5.4, group E, nos. 6, 7). In the classes Miliolata, Nodosariata, and Rotaliata the whole multichambered shell could be entirely enveloped by the last chamber (Fig. 5.4, group E, nos. 1, 3, 9), looking from the exterior like the unilocular shell. In such cases there is a return to the previous unilocular state but on the base of a new more advanced multichambered state (new level of the developmental spire). In rare cases such oligomerization of the chambers within a single foraminiferal organism happens as a result of dissolution of the inner chamber wall and fusion of several chambers into a single one. This phenomenon occurs in Foraminifera in the class Nodosariata (e.g., *Bombulina*; Fig. 5.4, group E, no. 5). The process of oligomerization also takes place in many multicellular groups, for instance, the fusion of the breast segments in Insecta and the reduction of the number of parapodia in Polycheta.

The differentiation of the apertural system of the foraminiferal shell into the main and supplementary ones was mentioned above.

The polymerized foraminiferal nuclei were the subject of subsequent morphological and functional differentiation into somatic and generative nuclei (Fig. 5.4, group D, nos. 4, 5) in the classes Miliolata and Rotaliata. The representatives of the class Spirillinata that were studied appeared to be homokariotic (Fig. 5.4, group D, no. 3). Data on the nuclear apparatus of Nodosariata are absent.

In Metazoa the vegetative and reproductive functions are shared between different organs. The processes of polymerization and differentiation at the nuclear level are also specific features of unicellular organisms. Polymerization of the nuclei is known in such protistan groups as Filosea, Testacealobosea, and Opalinata. But the polymerized nuclear apparatus underwent the next stage of evolutionary development, namely, the differentiation of the nuclei in terms of form and function into somatic and generative ones is known only in the two large and advanced protistan groups: in Infusoria and Foraminifera. This fact along with the other advanced structures of the organisms, such as, for instance, a very complex and differentiated foraminiferal calcareous skeleton or an infusorian cytoskeleton, provides a basis for regarding both groups as taxa of a high taxonomic rank—as separate phyla (Mikhalevich 1980, 1981, 2000). During foraminiferal development the degree of polymerization and differentiation of the nuclear apparatus increased from the more primitive classes (Astrorhizata, Spirillinata) to the more advanced ones (Miliolata and Rotaliata). A similar parallel complication is accompanied by the development of the skeletal structures (Fig. 5.2).

5.4 Processes of Integration in Foraminiferal Development

In polymerized and differentiated shells, especially in the supermultichambered forms, the communication of the initial chambers with the final ones and with the environment becomes hard. The equilibrium between the organism and the surroundings is broken and the organism comes to an unbalanced state (Prigozhin and Stingers 1986). In this case the rise of the new integrative system facilitating the communication between the disintegrated parts of the organism reinstates the lost equilibrium state of the organism. This third stage (integration) is represented in the foraminiferal shells by such integrative systems as foramens, tunnels, stolons, apertural integrative systems, and systems of canals (Fig. 5.5).

The simplest of all of these systems is the system of the inner foramens between the chambers which are formed when the outer main aperture turns out inside the shell after the formation of a new chamber above the previous one during the growth process (Fig. 5.5, nos. 1–6). Such foramens occur in all four classes with multichambered shells. They permit cytoplasmic flows from chamber to chamber and migration of the nuclei along the shell during the reproductive processes (Fig. 5.4, group D, no. 6). In some more advanced genera in addition to these inner foramens new secondly formed systems of passages exist. These are tunnels and stolons. Tunnels are known in Paleozoic fusulinids, which are regarded in the new system as belonging to the class Miliolata (Mikhalevich 2004, 2006, unpublished results). They

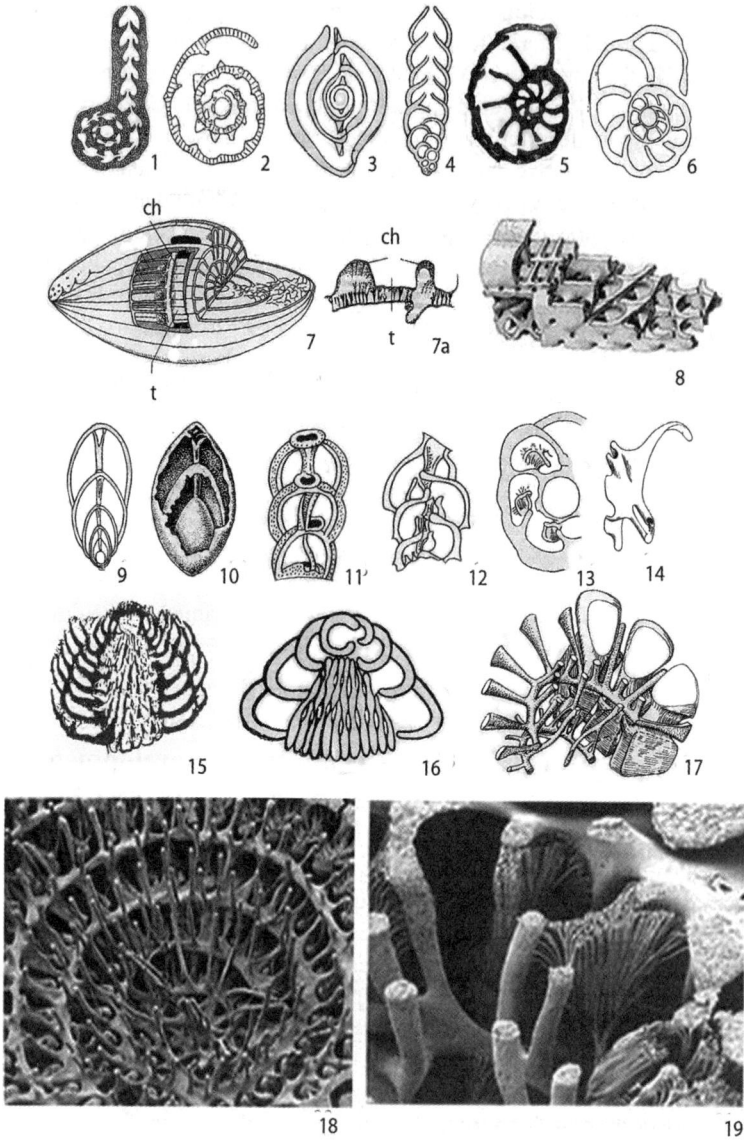

Fig. 5.5 Integrative systems in the foraminiferal shell (not to scale). 1–6 foramens between the chambers: *Ammobaculites* (1), *Melathrokerion* (2), *Pyrgo* (3), *Bigenerina* (4), *Lenticulina* (5), *Cibicides* (6). 7 chomata (*ch*) and tunnels (*t*) in Fusulinoida (7 the whole shell, 7a the detail). 8 system of stolons of *Marginopora*. 9–14 inner apertural integrative systems: *Ellipsoglandulina* (9), *Ellipsoidina* (10), *Siphogenerinoides* (11), *Eouvigerina* (12), *Reincholdella* (13 chambers with intercameral plates, 14 intercameral plate, enlarged); 9, 10 Nodosariata; 11–14 Rotaliata). 15–19 canal systems: *Lasiodiscus* (15), system of fissures in the center of the shell, *Lockhartia* (16), system of fissures between the pillars in the center of the shell, *Pseudorotalia* (17), system of tubes going between the chambers, *Elphidium* (18), *Pararotalia* (19); 18, 19 chambers are dissolved, tubes of the canals clearly seen. 15 Spirillinata; 16–19 Rotaliata. (8 after Hottinger 1978; 9–14 after Mikhalevich 2000, modified; 15 after Rauzer-Chernousova and Fursenko 1959; 16 after Sigal 1956; 17-19 after Hottinger 1979)

represent the space between the chomata—the ridgelike secondary deposits of the shell matter inside the chambers at their bottom (Fig. 5.5, no. 7a). Tunnels permit and direct in an easier way the cytoplasmic flows. The stolon system is the system of passages between the chambers, studied in detail by Hottinger (1978) (Fig. 5.5, no. 8). It is known among the representatives of the classes Miliolata and Rotaliata. Those three systems are rather primitive and do not have their own walls limiting their space.

Integrative apertural systems are developed in the classes Nodosariata and Rotaliata. They represent the inner apertural structures connected from chamber to chamber into one joined system and going inside the chamber space. In Nodosariata they are formed by the closed entosolenian tubes (*Ellipsoidina, Ellipsoglandulina, Pleurostomella*) (Fig. 5.5, nos. 9, 10). In Rotaliata they could be represented also by closed tubes (though of a character different from that in Nodosariata) or more often by half-closed tubes, grooves, tooth plates, or complexes of tooth plates of variable form (Fig. 5.5, nos. 11–14). These integrative inner apertural systems partly or fully isolate cytoplasmic flows and serve as supporting structures inside the apertures where the cytoplasmic flows are more intense. They provide full or partial differentiation of the cytoplasm inside the cell and also help to canalize the cytoplasmic streaming.

The most complex integrative system in Foraminifera is the system of canals having its own walls and going not inside but between the chambers. These systems were studied and demonstrated in the intricate works of Hottinger (1979; Fig. 5.5, nos. 17–19). Complex canal systems are developed in many higher representatives of the class Rotaliata when there are thin canal tubes between the chambers in spiral and radial directions often ramifying into the complex system of multiple thinnest tubes (Fig. 5.5, no. 19). Canal systems of the higher representatives of the class Spirillinata are less complex and differ from those of the higher Rotaliata: their tubes go between the whorls of one tubular chamber rather than between the multiple chambers of Rotaliata; their branching is not so complex or represents the system of fissures (Fig. 5.5, no. 15). In the Rotaliata themselves some more primitive "canal" systems exist, formed by the system of radial and intraseptal fissures and interlocular spaces between the chambers, facilitating their communication with the environment as they are opened into the wide and partly opened umbilical area (*Ammonia, Lockhartia*; Fig. 5.5, no. 16).The space of these passages is separated by secondary lamellae or pillars rather than by the closed tubes of the more advanced Rotaliata representatives. The early chambers of such shells often do not even have such an integrative system of passages and are more strongly isolated. In Nodosariata a specific canal system arose only once as an exception in one genus *Delosina* (Revets 1989) and is represented by a system of thin tubes going under the sutures and opening along them.

Hottinger (1978) showed that the ectoplasm circulates in canals supplying the cell with environmental elements, including oxygen. Thus, forming the complex ramifying system of thin and thinnest hard calcareous tubes, it plays the role of the integrative blood system in higher multicellular animals and its complexity is striking for a unicellular organism. All the last three types of foraminiferal integrative systems (stolons, apertural integrative systems, canals) represent the systems of the new organism level in one cell, thus giving at the same time an example of the multiplication (polymerization) of the organism systems. In some genera the two types of integrative systems could be combined in one shell (for instance, the stolon system coexists with the canal system in *Nummulites* and in some other nummulitid genera). The primitive system of foramens between the chambers also always occurs in all the shells possessing the more complex integrative systems.

Each integrative system has its own material object. The material objects of tunnels are represented by chomata, stolons—by a rather complex system of alternating openings, and canals—by the system of fissures limited by the secondary lamellae in the less advanced forms and by the closed tubes in the highly advanced genera. In the case of canals their inner ectoplasmic content also represents their special material object. Similarly in the metazoan blood system the vessel's walls are composed of specialized cells and their content consists of the different types of blood cells. Of course the blood system is a much more complex plural-component and more deeply differentiated system than the foraminiferal canal system developed within one cell. But all such integrative systems permit the direction and canalization of special flows providing their special functions, thus facilitating the communication of the different parts of organisms and increasing their wholeness. They provide the interaction of the disconnected parts and the strengthening and intensification of such interaction.

5.5 Aromorphoses in Foraminiferal Evolutionary Development

With the appearance of such a complex integrative system as the canal system the foraminiferal organism achieved a new higher level of its organization, giving it new and more effective possibilities of surviving and competing with other organisms. Such significant progressive promising evolutionary changes were named by Severtzov (1925, 1939) "aromorphoses." Later A. Panov (personal communication) called them "phase transfer," moving the term nearer to the physical notions.

In the foraminiferal evolutionary development the rise of multichamberedness and of the complex integrative canal system are considered as such aromorphoses (Mikhalevich 1981, 2000). Among the types of secreted calcareous shell walls, microgranular and porcellaneous wall types are composed of disordered randomly oriented crystals (Fig. 5.4, group F, nos. 1, 2). In Spirillinata, Nodosariata, and Rotaliata their calcareous crystals are strictly oriented perpendicularly to the shell surface (Fig. 5.4, group F, nos. 3, 4). Such crystal disposition makes the shells look hyaline and often translucent. But only in the latter of these three classes are these perpendicularly organized crystals situated in two lamellae with the middle organic layer between (Fig. 5.4, group F, no. 4). In this case the wall layer is doubled in the process of polymerization. The functions of these two layers (lamellae) also differ, which can be considered as a result of the process of differentiation. The ultrastructural disposition of the calcareous crystals in the foraminiferal shell wall serves as the material object of the shell construction. It is possible to regard the rise of the bilamellar character of the calcareous secreted shell wall of the higher Rotaliata as aromorphoses also. This type of shell ultrastructure provides a less heavy and thin wall possessing new constructive possibilities. As a result Rotaliata have three additional aromorphoses in their development compared with the class Astrorhizata and two additional aromorphoses compared with the other classes. This gave them the opportunity to develop during their geological history the most complex shell structures with the greatest degree of functional differentiation and specialization of the parts (e.g., floating and brood chambers, canals). All this in turn provided them with a new level of organization, new complexity and wholeness of the organism, higher evolutionary rates of development and new possibilities in their competition with the other groups in the environment, adaptive radiation, and expansion into new ecological niches (Fig. 5.3). Rotaliata is the only foraminiferal group which could transform from the bottom to the pelagic mode of life really getting up from the bottom.

Aromorphoses provide the organisms with the transition to a new level of organization which usually goes through the stage of a nonequilibrium, unbalanced state (Prigozhin and Stingers 1986) (see earlier). The creation of the new integrative system permits the organism to pass such an unbalanced state successfully and to preserve and enlarge its wholeness. The level of organization increases.

As described above, the processes of polymerization (with the subsequent differentiation) and integration and the aromorphoses happened in different classes of Foraminifera independently and in parallel, having in each of them similar tendencies and at the same time their own specific features.

5.6 General Character of the Main Evolutionary Regularities

5.6.1 Different Levels of the Organization of Matter

This chapter does not regard the other important regularities as contributors to the evolutionary process, such as block-modular principles, feedback, nonlinearity, adaptations, increase of the rate of the evolutionary change, adaptive radiation, divergence, convergence, and parallelism, symbiotic evolution, and some others, but only the main stream ways—polymerization, differentiation, integration, and the phenomenon of aromorphoses. All of them give the organism a higher degree of wholeness, activation, and mobilization of its functions owing to the increased reciprocal activity of its functional parts and their better interaction.

These rules were significantly elaborated by the Russian evolutionary school (Berg 1922, 1977; Dogiel 1929, 1954; Severtsov 1925, 1939; Schmalchgausen 1946; Beklemishev 1964; Podlipaev et al. 1974; Naumov et al. 1977; Golubowski 1994; Poljanskiy and Raikov 1977, and others) mostly for the Metazoa. Here I try to demonstrate the manifestation of these developing processes using as an example one of the groups at the unicellular level. The continuum of the evolutionary processes of inorganic and living organic matter (biolevel) which represents the more complex level of organization and the generality of the rules of the development for both levels were shown in Krylov and Libenson (2002). They put the "mark of equation" between the biogeochemical succession and the evolution of the biosphere. Wide character of the evolutionary regularities and the possibility of their application to the lower levels of organization of matter, beginning from the physical and chemical levels, was shown in a set of publications (Larin 1977; Krylov and Libenson 2002; Libenson and Przhibelskii 2003; Krylov 2005). Social life represents the most complex higher level of the biological form of life and at the same time the highest level of cosmic life. The processes of polymerization, differentiation, and integration along with the phenomenon of aromorphoses being the general universal evolutionary rules could be applicable to human social life as well.

5.6.2 Processes of Polymerization, Differentiation, and Integration in the Development of Human Society

Let us have a brief look at the history of human society. It developed through the shoal to the genus, tribe, then from the union of tribes to the state. The number of individuals increased in the human genera and tribes (polymerization), then more significantly in the union of tribes (new level of polymerization). (This process still is continuing and mankind has now reached a population of about six billion.)

The enlargement of the society of ancient human beings and the rise of more complex social relations in the process of collective joint actions (e.g., hunting, defense against wild animals) resulted in the necessity for a communication system, of the canalization of these new relations. It was a question of survival, of integration—and language emerged, the first social integrative system of human society. For Cro-Magnon man, not being very strong physically, this was of more importance than for other human species. The rise of language gave human beings an advantage during collective actions, quicker adaptation, and regulation. It provided the interaction of the disconnected parts and the strengthening and intensification of such interaction. Even in the early human stages of development from the shoal to the genus and tribe some elements of differentiation could be marked, the most primitive of them was the division into age groups and division of labor into man's and woman's work. The further differentiation of labor in the tribe and unions of tribes caused the appearance of the three social groups: laborers, warriors, and priests. There were heralds to transfer the messages from tribe to tribe as the separate tribes needed communication between them just as the primitive multichambered foraminifera needed their foramens for communication between their chambers. The functions of such groups in ancient society differed. They were accordingly the creation of material products, security, and preservation of knowledge and information. Heralds provided communication between the tribes (prototype of the future postal service). With the advent of priests the division into physical and mental work occurred—the most important event in the evolution of human social life (Achnazarov 2002). Complex tasks in the process of joint works such as growing crops or building dams to create and enlarge the economic surplus (the necessary circumstance permitting a society to keep and feed priests and warriors) demanded more complex structures and relations in the society for it to survive.

When the amorphous tribal society became differentiated into different social groups a new integrative system emerged—there was a state and relations governed by a system of law and legal regulations. This is an example of centralized structures overcoming dissociation. The origin of the state could be considered at the same time as an event of aromorphic character. If, in the union of tribes, the chieftain was the supreme power (with the variants—sometimes he was simultaneously the leader of the bodyguard) while the other functions of management (collection of taxes, realization of justice, and many others) were not differentiated and were performed by the council of elders in the state during the processes of centralization (i.e., integration), the main part of the latter functions passed to the king, and the counts and barons preserved only the right of possession of their land. The functions of management became differentiated: special positions (and later special institutions) appeared for their execution.

The following is also one of the evolutionary rules: the limitation of the functions of the subordinate structures in favor of a higher integrative structure (the subordination of the separate foraminiferal chambers to their unicellular organism as a whole, and of the metazoan cells to the centralized multicellular organism).

Though the comparison of the state with the human organism was mentioned earlier, it had a more artistic image and likening based on domestic notions rather than on an analytical approach based on scientific knowledge. Further it did not bring to light the mechanism of development of such a similarity. In such a likening the specific differences and complexity of each of these systems existing at the different levels of their organization were not considered.

State and law systems developed from the more primitive archaic eastern monarchy and medieval monarchy to the early bourgeois society of the new time and then to the industrial and postindustrial societies of the newest time and lastly to the information society.

In the medieval state the differentiation of the previously existing social groups was continued, mainly in relation to the different kinds of human activity (e.g., multiple guilds, corporations, etc.)—"the blooming complexity" (a term used by the well-known Russian philosopher. Leontiev 1996) of the Middle Ages. The partitions between these differentiated groups were strict: belonging to a definite group was defined by origin and usually could not be changed (as an exclusion only).

With the rise of bourgeois society centralization and integration increased. Gradually the medieval fragmentation and disunity was overcome; strict partitions between the different social groups were broken. Belonging to different social groups was not defined anymore by origin, but was mostly a result of personal possibilities and achievements, and differentiation of society was carried out on account of the more specific distinctions (educational, on the basis of property, etc.). The next very important stage of differentiation of the higher integrative functions of the state was the transition of the functions of the king's personal particular order to sectoral management by means of the institutions (the beginnings of ministries), and separation of the functions of the supreme power from the functions of management.

The disappearance in the bourgeois state of the estate partitions existing before could be compared with the disappearance of the inner chamber wall (dissolution of inner septa) in some foraminifera such as in *Bombulina* (Fig. 5.4, group E, no. 5) and the secondary reestablishment of the unilocular stage of the organism at the new level of organization. In both cases the elimination of the partitions made communication of the different parts easier and created in both systems a more dynamic state. The more effective social dynamics resulted in a more developed and effective economy, thus providing to such a state advantages in international competition.

The transition from the feudal to the bourgeois state also had an aromorphic character as later did the process of transition to the recent postindustrial and information society and state. This process is not yet accomplished. This type of society consists of a big number of groups where people are differentiated and organized on the new principles mainly on the basis of their interests (not only material or industrial, but mostly wider and more variable: societies of fishermen, lovers of art, football fans, etc.). These groups are connected in the more complex and principally new systems of intercrossing and multiaspect connections resembling the human brain more with its neuron net than the blood system and forming the new "blooming complexity" at the higher level of organization.

The new process of recent integration is represented in international suprastate structures (e.g., the European Union, currently including 27 countries, international currency reserves, the World Bank), in the processes of globalization. These suprastate structures resemble the integrative stage of the unions of tribes but also represent a new higher and more complex level of organization.

The integrative law systems of the different types of states mentioned above developed correspondingly from the primitive feudal law to the civil law after the Magna Charta and the Napoleonic Codex.

It is possible to compare very early human society not differentiated into social groups with the primitive unilocular foraminifera not divided into chambers, and the law systems of the states of different levels of their organization with the different types of the foraminiferal integrative systems in their different taxa with the multichambered shells existing at various stages of their development and complexity. Thus, feudal law can be compared with the more primitive integrative systems of the foraminiferal shell such as tunnels and stolons (Fig. 5.5, nos. 7, 8), and the later civil law with the perfect canal system of the higher Rotaliata (Fig. 5.5, nos. 17–19). The law systems during their further development diverged into two branches which can be outlined as codified common law (based on codex—German-Romanic branch) and based on court decision (Anglo-Saxon for all four branches of law). The latter can be compared with the most advanced Rotaliata canal system'; the first one with the variant of fissures (Fig. 5.5, nos. 15, 16). A society composed of different groups cannot exist without a system of law in a similar way to how the supermultichambered foraminiferal shells cannot survive without their integrative systems which provide easier cytoplasmic flows. Of course, the structures of societies with their multiple and interconnected variational components are many times more complex than the structures of one organism at the unicellular level. But the mechanisms of their development through the processes of polymerization, differentiation, and integration and via the nonequilibrium state and new aromorphic changes to the new organizational level are the same.

The law systems are not the only integrative systems of human society. Moral and religious standards also serve as such systems as the laws do not embrace all aspects of life in human society. Each of these standards acts in its own sphere (has its special functions). Economics and business are based on legal relations. Morals represent a more ancient system (compare this with the system of simple foramens in foraminifera; Fig. 5.5, nos. 1–6) which appeals to the simpler and more natural spheres of being. Religion can serve as an integrative system inside the state and as a system of the new civilization level uniting several states (Christian civilization).

5.6.3 Processes of Polymerization, Differentiation, and Aromorphoses in the Development of the Material Objects of Social Integrative Systems

Like the foraminiferal canal system and the Vertabrata blood system and like the brain for language each social integrative system also has its own material object. For the state this includes bureaucratic institutions, and for the legal system law institutions (Raskin 2001). Different cultural institutions serve as a base for different social groups of the postindustrial state. Communication service is the material object for several integrative systems at once. The evolutionary development of material objects (of different types of integrative systems) also passed through historical stages of polymerization, differentiation, integration, and aromorphoses. Whatever example is used all of them display the same regularities.

The first human production tool was a primitive stone cutting instrument; later a multitude of instruments were used, more complex, multipurposed, and specialized (nomenclature of the Late Stone Age counts dozens of them). The advent of metal instruments (firstly made from copper and bronze and later from iron) could be regarded as aromorphoses. The hand-manufactured production in the guilds of medieval times underwent a similar process of multiplication and differentiation until it was substituted by machine production. This new mode of production also represented an aromorphic change. Polymerization of the bearers could be exemplified by the polymerization in variable technical branches—multibarrel arms (revolvers), multiengine airplanes, etc.

One of the most important methods among these material objects for the integration at the social human level is the communication service which serves as a material object for several integrative systems. The line of its stages can be seen from the first heralds and sound signals to the regular postal service (relay-race horse mail, written mail) and from communications based on the wire principle (telegraph, telephone) to wireless ones based on wave principles (radio, mobile phones). In parallel the rise and development of printing and an increase in book and press publications took place. It was not by chance that the dissemination of Protestantism coincided with widespread book-printing. With the appearance of television and the Internet and with the integration of all the communication services (television, Internet, fixed and mobile communication systems) into a unified whole there was a unified information system. The Internet represents a new type of communication that is difficult to distinguish from material matter and the integrative system itself. The recent units of countries would not be possible without a new unified communication service system as its material objects.

In the development of material objects the examples of aromorphic changes also happened many times (invention of firearms, wireless communication service). The most demonstrative example would be the transition from the multicore processor to the quantum computer.

With the increasing complexity of human society, the number of integrative systems increases. While in the foraminiferal organism the maximum number exceeds only four, in the human organism it is about ten, and the number in human society increases dozens of times.

5.7 Conclusions

We have shown that foraminiferal multichamberedness and the different types of calcareously secreted walls of the integrative systems originated in different classes many times independently and in parallel, and not simultaneously (e.g., canals of Lasiodiscids in the class Spirillinata in the Paleozoic, and of Rotaliata much later, since the Upper Cretaceous).

The origin of the states had the same character, often with the existence of intermediate forms (between the feudal and bourgeois, or the early feudal and late feudal societies). The ancient primitive structures and the advanced ones usually coexist simultaneously (unilocular Foraminifera live in the Recent seas together with the Rotaliata, i.e., unicellular organisms coexist with multicellular ones; the feudal type of state coexists with bourgeois and industrial societies). Human tribes existing at the level of the Stone Age still exist in South America, where they are not even acquainted with making fire. And the newly formed structures also often contain the remnants of the previous one: some remnants of the feudal law in civil law, of the estate partitions in bourgeois society (the caste system in modern India and the House of Lords in the UK have been preserved, for example).

The transition from one organizational level to a new one occurs when the possibilities of the previous structures are exhausted. It happens as a result of aromorphoses and usually goes through a nonequilibrium stage. Thus, the eukaryotic Rotaliata cells of higher complexity represent the highest level of organization at the unicellular level and the mainstream of evolution of life later went the way of multicellularity in a similar way as the previous Procariota level was followed by the eukaryotic way. When the potentialities of the feudal-level state were exhausted bourgeois societies arose. The material objects of human activities exemplify the same regularities. Stone tools were replaced by metal tools, hand manufacturing by factory production, mechanical principles of arms and aircraft construction by the jet principle, etc.. In materials bearers of foraminiferal integrative systems chomata and stolons were replaced by canals. One of the best examples is seen by the changes in Japan at the end of the nineteenth century (Maidzy revolution) when the elimination of the estate compartmentalization (of saoguns) resulted in equal rights and the advent of a new effective integrative systems of the western type (including the judiciary). These were enough to make the leap in development through several centuries at once.

The recent democratic rightful state having the most advances integrative systems with the most developed material objects of these systems represents the highest level of social organization, which gives it the most effective possibilities for surviving in competition with other state systems. The foraminiferal higher class Rotaliata was not only widely distributed in at the ocean bottom but owing to its highly developed integrative systems and to such a progressive material object of its shell as the bilamellar calcareously secreted wall it was able to gain a new ecological niche and to expand into pelagic waters. Similarly the information society of the new type possesses social systems with a high level of integration and also highly organized material objects. All this gives mankind possibilities to expand into the cosmos.

This short historical discourse was made to search for the mechanisms of evolutionary changes at different levels of structural organization of organic and social life and to show the continuum of these life levels and the general character of the main evolutionary regularities (polymerization, differentiation, integration, and aromorphic changes) through time. The same mechanisms could be applied and searched for within different organizational levels of varying complexity. There was no opportunity here to go into more detailed analysis of the peculiarities of the processes of polymerization and differentiation themselves such as polymerization through the subdivision of homologous units or through their new formation and addition, and cases of differentiation after oligomerization (through the process of fusion or elimination of subunits, etc.) and only the main processes could be considered. Knowledge of all of these regularities permits not only an understanding of the evolutionary processes of the past and present but also the ability to predict ways of development for the future.

Thus, the basic evolutionary regularities elaborated first for living organisms represent general evolutionary principles applied to all levels of organization of the matter of the universe. They promote the mechanism of transition to a new level of organization and the way of further progressive evolutionary changes through time.

References

Achnazarov EB (2002) Kontury evolucii (The counters of evolution). Nedru, St. Petersburg
Beklemishev VN (1964) Osnovy sravnitelnoj anatomii bespozvonochnych, 3rd edn (Fundamentals of the comparative analysis of Invertebrata). LA Zenkevich (ed) Akademiya Nauk SSSR, otdelenie biologicheskich nauk. Nauka, Moscow

Berg LS (1922) Nomogenez ili evoluciya na osnove zakonomernostej (Nomogenez or the evolution based on regularities). St Petersburg
Berg LS (1977) Trudy po teorii evolucii, 1922–1930. Nauka, Leningrad
Cuvier G (1801) Lecons d'anatomie comparee. Recuellies et publ. sous ses yeux par G. Dumeril, vol 1, Paris
Darwin C (1859) The origin of species by means of natural selection, or the preservation of favoured races in the struggle of life. John Murray, London
Dogiel VA (1929) Polymerization als ein Prinzip der progressiven Entwicklung bei Protozoen. Biol Zentralbl 49:451–469
Dogiel VA (1954) Oligomerizacija gomologichnych organov kak odin iz glavnych putej evolutcii zhivotnych (Oligomerization of the homologous organs as one of the main ways of the animal evolution). A.A. Strelkov (ed.) Izdatelstvo Leningradskogo Universiteta, Leningrad
Golubowski MD (1994) Classicheskaya I sovremennaja genetika: evoluciya vzgljadov na nasledstvennuju izmenchivost (Classical and modern genetics: evolution of views of hereditary variation). Tr Spb Ova Estestvoispyt 90:37–48
Hengeveld R, Fedonkin MA (2004) Causes and consequences of eukaryotization through mutualistic endosymbiosis and compartmentalization. Acta Biotheor 52(2):105–154
Hottinger L (1978) Comparative anatomy of elementary shell structures in selected larger Foraminifera. In: Foraminifera. R.H. Hedley (ed) C.G. Adams. Academic Press, London, New York, San Francisco Vol 3.1: 203–206
Hottinger L (1979) Araldit als Helfer in der Mikropalaeontologie. Aspekte. Giba-Geigy AG. Division Kunststoffe und Additive, Nachdruck von Wort und Bild mit Quellenangabe gestattet. Uber nur mit Genehmigung der Redaktion, Basel 3: 1–10
Krylov MV (2005) Initial principles of the evolution of living matter. Proc Zool Inst Russ Acad Sci 308:35–40
Krylov MV, Libenson MN (2002) Continium evolucionnych processov zhivoj I nezhivoj materii (Continium of the evolutionary processes of the organic and inorganic matter). In: MN Libenson (Chairman of the Programme Committee) Prblemy i perspectivy mezhdisciplinarnych fundamentalnych issledovanij. Materials of the second scientific conference of S. Petersburg Association Scientists and Scolars (SPASS), 11–12 April 2002. Sankt-Peterburgskiy Sojuz Uchenych, S. Petersburg, p 55
Lamarque JB (1822) Histoire naturelle des animaux sans vertebras, vol 7. Lamarque, Paris
Larin JS (1977) Proischozhdenie i zhiznedejatelnost kletochnoj organizacii v svete idej polymerizacii i oligomerizacii V. A. Dogielja (The origin and vital activity of the cellular organization in the light of the ideas of the polymerization and oligomerization of V. A. Dogiel). In: Significance of the processes of polymerization and oligomerization in evolution]. In: OA Scarlato (ed) Published by Zoological Institute of the Academy of Sciences USSR, Leningrad, pp 89–93.
Leontiev K (1996) Vostok, Rossiya I slavjanstvo (The East, Russia and Slavdom). Respublika, Moscow
Libenson MN, Przhibelskii SG (2003) Concepciya sovremennogo estestvoznaniya (Concepts of modern natural science. Part II. Physics). International Banking Institute, St Petersburg
Loeblich AR Jr, Tappan H (1964) Treatise on invertebrate paleontology. Part C. Protista. 2. Sarcodina, chiefly "Thecamoebians" and Foraminiferida. In: Moore RC (ed).: Geological Society of America and University of Kansas Press, Lawrence. Vol. 1, 2. References: 797–868
Loeblich AR Jr, Tappan H (1987) Foraminiferal genera and their classification. Van Nostrand Reinhold, New York
Mikhalevich VI (1980) Sistematika i evolyuciya foraminifer v svete novyikh dannyikh po ih citologii i ul'trastrukture (Systematics and evolution of the Foraminifera in view of the new data on their cytology and ultrastructure). In: Principy postroeniya makrosistemy odnokletochnyih zhivotnyh. (The principles of the formation of the macrosystem of the unicellular animals). Tr Zool Inst Akad Nauk SSSR 94:42–61

Mikhalevich VI (1981) Parallelism i konvergencia v evolucii skeletov Foraminifer (The parallelism and convergence in the evolution of the Foraminiferal skeleton). Tr Zool Inst Akad Nauk SSSR 107:19–41

Mikhalevich VI (1992) Makrosistema foraminifer (The macrosystem of the Foraminifera). Zoological Institute Russian Academy of Sciences. Doctoral thesis, St Petersburg

Mikhalevich VI (1998) Makrosistema foraminifer (The macrosystem of the Foraminifera). Izv Ross Akad Nauk. Ser Biol 2:266-271

Mikhalevich VI (1999) Sistema i philogenia Foraminifer (The foraminiferal system and phylogeny). Biol Ser Vyp 3(1):11–30

Mikhalevich VI (2000) Tip Foraminifera d'Orbigny, 1826—Foraminifery (The phylum Foraminifera d'Orbigny, 1826–Foraminifers). In: Alimov AF (ed) Protisty: Rukovodstvo po Zoologii, part 1. Nauka, St Petersburg, pp 533–623

Mikhalevich VI (2004) On the heterogeneity of the former Textulariina (Foraminifera). In: Bubik (Czech Geol. Survey, Brno Branch, Leitnerova 22, Brno, Czech Republic), M, Kaminski MA (Department of Earth Sci., University Colleague London, Gower Street, London, WCIE GBT, U.K.) (eds). Proceedings of the Sixth international workshop on agglutinated Foraminifera(Prague, Czech Republic, September 1–7, 2001). Published by Grzybowski Foundation special publication, vol. 8.Printed in Poland by Drukarnia Narodova, Krakow) pp 317–349

Mikhalevich VI (2005) Polymerization and oligomerization in foraminiferal evolution. Stud Geol Pol 24:117–141

Mikhalevich VI (2006) The similarity of Fusulinoida and Milioloida—convergence or parallelism? Proc Zool Inst Russ Acad Sci 310:133–137

Mikhalevich VI, Debenay J-P (2001) The main morphological trends in the development of the foraminiferal aperture and their taxonomic significance. J Micropalaeontol 20:13–28

Naumov AD, Borkin LJ, Podlipaev SA (1977) Principy polymerizacii i oligomerizacii: processy i systemy (Principles of polymerization and oligomerization: processes and systems). In: Znachenie processov polimerizacii i oligomerizacii v evolucii (Significance of the processes of polymerization and oligomerization in evolution). OA Scarlato (ed) Published by Zoological Institute of the Academy of Sciences USSR, Leningrad, pp 5–8

Pawlowski J, Holzmann M, Bemey C, Fahrni J, Gooday AJ, Cedhagen C, Habura A, Bowser SS (2003) The evolution of early Foraminifera. Proc Natl Acad Sci USA 100(20):11494–11498.

Podlipaev SA, Naumov AD, Borkin L J (1974) K opredeleniju ponjatij polimerizacii i oligomerizacii (On the definition of the notions of polymerization and oligomerization). Zh Obstsch Biol35(1):100–113.

Poljanskiy Ju, Raikov IB (1977) Polimerizacija i oligomerizacija v evolucii prostejshich. (Polymerization and oligomerization in the evolution of protists). In: Znachenie processov polimerizacii i oligomerizacii v evolucii (The role of the precosses of polymerization and oligomerization in evolution). Leningrad, pp 29–32

Prigozhin I, Stingers I (1986) Porjadok iz chaosa (Order out of chaos). Progress, Moscow

Raskin R (2001) Rossijskaja imperija XIXnachala XX veka kak systema gosudarstvennych uchrezhdenij, sluzhby, soslovij, gosudarstvennogo obrazovaniya I elementov grazhdanskogo obstschestva (Russian empire of XIXbeginning of XX centuries like a system of the state institutions, service, classes, state education and the elements of the civil society). Mellen, Lewiston

Rauzer-Chernousova DM, Fursenko AV (1959) Osnovy paleontologii. Obshcaya chast', Prosteyshie (Principles of paleontology. General part and Protozoa). JuA Orlov (ed) Izdatelstvo Academii Nauk SSSR, Moscow

Revets SA (1989) Structure and taxonomy of the genus Delosina Wiesner, 1931 (Protozoa: Foraminifera). Bull Br Mus(Nat Hist) Zool Ser 55:1–9

Schmalchgausen II (1939) Puti i zakonomernosti evolucionnogo processa. Izbrannye trudy. (Ways and regularities of the evolutionary process. Selected works.).. AN SSSR, Sekcia chimikotechnologicheskich i biligicheskich nauk In-t evolucionnoy morphologii i ekologii zhivotnych im. A.N. Severcova, Nauka, Moscow

Schmalchgausen II (1946) Factory evolucii (Factors of evolution). Izdatelstvo Academii Nauk SSSR, Moscow-Leningrad

Schultze M.S. 1854. Ueber der Organismus der Polythalamien (Foraminiferen) nebst Bemerkungen ueber die Rhizopoden im Allgemeinen. Engelmann, Leipzig

Severtzov AN (1925) Glavnye napravlenija evolucionnogo processa. Progress, regress i adaptacii (Main directions of the evolutionary process. Progress, Regress and Adaptation). Izdanie Tovaristschestva AV Dushnov i K°, Moscow

Severtzov AN (1939) Morphologicheskie zakonomernosti evolucii (Morphological regularities of evolution). BS Matveev and SV Emelianov (ed-s) Izdatelstvo Academii Nauk SSSR, Moscow-Leningrad

Sigal G (1956) Otrjad Foraminifery (Foraminifera). NN Subbotina (ed.) Gosudarstvennoe nauchno-technicheskoe izdatelstvo neftjanoy i gorno-toplivnoy literatury, Leningradskoe otdelenie, Leningrad

Voronova MN, Mikhalevich VI (1985) Sovremenye predstavlenija o zhiznennych cyclach foraminifer (Recent concept of the foraminiferal life cycles). Proc Zool Inst Acad Sci USSR 129:48–66

Zech L (1964) Zytochemische Messungen an den Zellkernen der Fora-miniferen Patellina corrugata und Rotaliella heterocariotica. Arch Protistenkd 107:295–330

Chapter 6
Old and New Concepts in EvoDevo

Margherita Raineri

Abstract The concept of the modular organization of the organism is central to both classical experimental embryology and modern evolutionary developmental biology (EvoDevo). The latter discipline often ascribes homology, or the diversity of forms arising from the same module or combination of modules, to gene mutations or changes of gene regulation. This gene-centred atomistic view does not yield reliable criteria for homology, nor can it account convincingly for phenomena such as convergence and cooption. To understand the logic behind this plasticity, attention should be shifted from variation of existing characters to the preconditions for homology. This could be done by identifying the basic modules and their molecular regulation within the context of the most general, shared embryonic morphology. Mapping divergent embryonic trajectories onto different fates of these modules may allow us to throw light on the constrained sets of different, but related morphologies which can arise from different combinations of modules.

6.1 Introduction

During the past two centuries the relationships between ontogeny and phylogeny have been the subject of extensive research and speculation. The most famous outcome of this intellectual enterprise is probably the fundamental biogenetic law (Haeckel 1866, 1874a), which claims that ontogeny recapitulates phylogeny and Animalia originated from an ancestor that was comparable to a gastrula formed by invagination of a hollow blastula. Its enduring popularity notwithstanding, the biogenetic law has always been controversial and at present seems to be definitely discredited (Richardson et al. 1997), yet, in recent times, the monophyly of Animalia has received consistent support from biomolecular studies showing unexpected

M. Raineri
Department of Biology, University of Genoa, Viale Benedetto XV 5, 16132 Genoa, Italy
raimrg@unige.it

conservation of the genetic networks from cnidarians to man (Conway Morris 2000, 2003; Erwin and Davidson 2002). The challenge of reconciling this molecular similarity with the amazing variation of biological forms has contributed to renewed interest in evolutionary developmental biology (EvoDevo). In spite of the benefits of unprecedented technical facilities, current attempts to elaborate an integrating theory linking developmental to evolutionary changes are thwarted by historically determined conceptual shortcomings. The present chapter analyses a few of these difficulties in order to suggest an alternative approach to the problem.

6.2 Two Different Approaches to Development and Evolution

Being regained under the influence of the neoDarwinian synthesis, the new alliance between ontogeny and phylogeny considers evolutionary changes to be dependent on selection acting on genetic changes. The promotion of this concept in the majority of the current literature has relegated somewhat to the fringe an alternative view suggesting that phenotypic modifications may be driven largely by epigenetic processes, in particular, variations of ontogenetic trajectories.

Pioneers of the latter approach were Geoffroy Saint-Hilaire (1818), and later on Albert von Kölliker (1864) with his theory of heterogeneous generation which ascribed the origin of novel body plans to changes occurring in the egg. For instance, according to von Kölliker an increase of yolk would affect the rates of growth of the different parts of the embryo, then the developing rudiments could be displaced and modified by the action of mechanical forces, and eventually phenotypic variation could be conspicuous, although real novelties could not appear, as every change would be the result of rearrangement and elaboration of elements which were already present in the ancestral forms.

While Haeckel's law is based on terminal addition yielding linear patterns of progressive developmental sequences (Fig. 6.1), heterogenesis makes reference to inherency and systemic modifications, implying neither a progressive trend, nor direct causation of character changes. Current studies supporting an origin of modern animals from complex ancestors and a developmental role of physical forces highlight the centrality of inherency and the importance of biomechanics to the evolutionary process (Albrecht-Buehler 1990; Galis and Sinervo 2002; Davidson and Erwin 2006; Stern 2006; Matus et al. 2006, 2007). However, as an offspring of nineteenth century biology von Kölliker's theory suffers from obvious limitations; in particular, it is concerned with spatial relationships of ideal structures, but it tells nothing of their differentiation. The link between these two aspects of ontogenesis was addressed by the newly born experimental embryology (Roux 1895) which paved the way to the theory of morphogen gradients and to the conceptions of induction and embryonic field, leading to painstaking experiments for identifying morphogens and inducers. Ironically, this strictly mechanistic approach only succeeded in demonstrating that the morphogenetic properties of the embryonic field could not be ascribed to any defined physical or chemical causal factor. In particular,

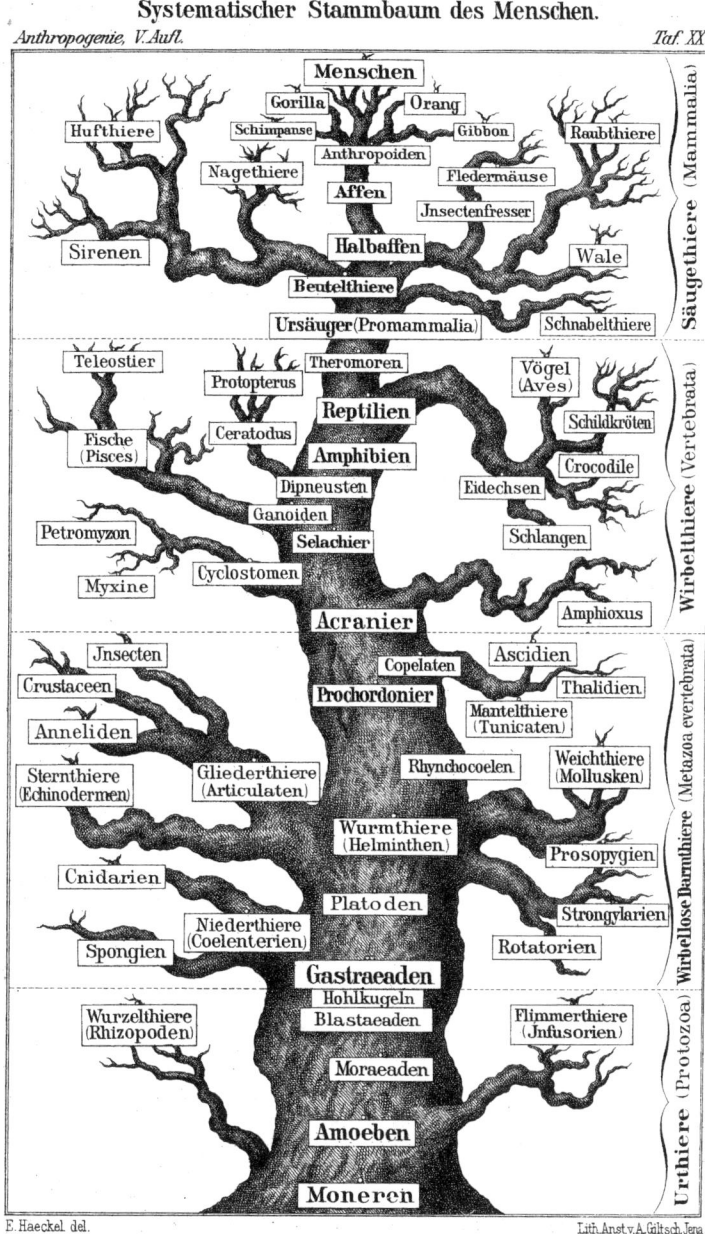

Fig. 6.1 The phylogenetic tree of vertebrates according to Haeckel, showing a progressive tendency of increased complexity towards the evolution of man. The lancelet (Acranier) is placed near the transition from invertebrates to vertebrates, but is considered to be a basal vertebrate rather than a protochordate. (From Haeckel 1874b, plate XX, courtesy of the Museum of Natural History "G. Doria", Genoa)

these properties appeared to include a spatiotemporal dimension and to exceed the limits of germ-layer and species specificity. Owing to this, Spemann (1938) came to the conclusion that 'the field by no means coincides, neither spatially nor logically, with the presumptive or with the determined rudiment of the organ originating in it' and 'induction is the same as field action, at least if one means by the latter, as I do, all sorts of agencies which may play a role in development'.

6.3 Homology, Modularity, Developmental Networks

Trying to realize how cell activities come to be coordinated in space and time so as to allow the harmonious development of the body parts, experimental embryologists have contributed to viewing ontogeny as the result of modular processes which can be combined in different ways depending on their reciprocal relationships (Gould 1977). Modularity has been rediscovered by the new EvoDevo also because it has a direct bearing on the concept of homology, which, in spite of all criticism and ambiguity of definition, is still essential to understanding the affinities between body plans and the role of ontogenetic variations in evolutionary changes (Raff 1996). The rudiments which according to Saint-Hilaire (1818) and von Kölliker (1864) keep their identity while being remoulded into different shapes could be the equivalent of both the classical and the modern genetically defined morphogenetic fields (Gilbert et al. 1996) and of the sets of characters or homologues which according to Müller and Wagner (1996) and Müller (2003) act as primary modular units of phenotypic organization and evolutionary transformation. Thus, a central point in the agenda of EvoDevo could be to unravel the relationships between homology and modularity, which may turn out to be the opposite side of the coin as to the links between spatiotemporal patterning and cell differentiation.

A basic difficulty in pursuing this goal is that dynamic natural entities such as modules and homologues can be defined only in the context of a process or a theory of a process, although there is no agreement on what this biological context could be (Wagner 1996). As the prevailing opinion tends to reduce this context to the genetic level, homologies are routinely investigated by comparing sequences and developmental expression patterns of genes. However, this comparative analysis meets with the lack of objective criteria for choosing representatives of different taxa and equivalent stages of diverging ontogenies, and moreover, with convergence at morphological and molecular levels, but conversely, dissociation between homologous structures and orthologous genes (Minelli 1998, 2003; McGhee 2000; Schierenberg 2001; Reed et al. 2007). With these caveats in mind, the comparison should be restricted to closely related taxa (Love and Raff 2006), although in this case the question of whether microevolution and macroevolution could depend on the same causal factors will remain unanswered, as the mechanisms underlying short-term divergence could not be extrapolated arbitrarily to higher levels of evolutionary change. Last but not least, this approach implies an atomistic view of biological forms since it is based on the assumption that each character should be controlled by a specific set of transcription-factor-encoding genes. As a consequence, the need for correlating the different parts of an organism into a functional morphological unit has led to ontogeny to be considered as the product of a 'developmental genetic programme' consisting of interconnected regulatory networks with a hierarchical organization, while phylogeny would be the result of changes of these networks hierarchically connected to structural variations and taxonomic levels (Davidson and Erwin 2006).

In this scenario, the arrow of time comes to be an essential constituent of the 'programme', as the temporal progression of development becomes translated into a hierarchy of networks corresponding, in turn, to a hierarchy of body parts. Such a linear mapping of genes onto phenotypes may be reminiscent of the model of the inverted cone of development (Arthur 1997), although according to other authors a developmental hourglass could be more appropriate as a metaphor, given the higher variability that can be detected both in early and in advanced stages (Duboule 1994; Raff 1996). In any case, this model raises a problem of level, for instance, does the specification of germ layers and heart involve the same hierarchical level of networks, as Davidson and Erwin (2006) seem to suggest? To consider fertilization to be the start of the programme taking into account only the temporal and genetic dimension cannot give an answer to this question. Fertilization triggers, for instance, the 'regulatory hierarchy of developmental genetic networks' that in sea urchin embryos is involved in endomesoderm specification (Davidson et al. 2002). However, the 'programme' involves a spatial dimension, because this genetic cascade is activated initially by asymmetrically distributed maternal determinants which bind to *cis*-regulatory modules. This maternal anisotropy is produced by many different factors, e.g. it includes a redox gradient which depends on a polarized distribution of mitochondria (Coffman et al. 2004), which, in turn, may be established by cytoskeletal patterns, and so on. At least in principle, changes of any of these parameters may lead to genetic changes, as *cis*-regulatory modules, even if they were 'static physical components of the programme' (Ben-Tabou de-Leon and Davidson 2006) could be processed in different ways by the cytoplasmic machinery acting not only on DNA sequences, but also on chromatin remodelling, which is essential for regulating the coordinated expression of genes (Meaburn and Misteli 2007).

Evidently, the developmental programme is not located as such in the genetic sequences, but may be coincident at the beginning with the genomic and cytoplasmic organization of the egg, and then with an intricate interplay between genetic and epigenetic regulation. Thus, conserved cell lineages and embryogenesis patterns may be determinant for bringing forth conserved *cis*-regulatory logic of gene expression in spite of high *cis*-regulatory sequence divergence, as shown, for instance, by the conserved regulation of *Otx* in ascidian embryos with divergent genomes (Oda-Ishii et al. 2005). Conversely, diverging ontogenies may establish comparable signalling, and therefore similar regulatory networks in different territories, for instance, dorsally in insects, but ventrally in vertebrates, as the result of inverted gradients of Sog/Chordin antagonizing Dpp/BMPs (De Robertis and Sasai 1996; Ashe and Levine 1999). According to Davidson and Erwin (2006) the expression of the same 'heart-field specification kernel' on the dorsal side of *Drosophila* and the ventral side of vertebrates may indicate that an ancestral heart gave origin during evolution to both arthropod and vertebrate hearts. However, this finding could be nothing but an example of the pervasive 'gene tinkering' (Jacob 1977) which follows necessarily from the limited range of developmental trajectories and the relatively low number and high conservation of the genetic circuits. If so, the key to deciphering phylogeny would be to discover not so much hierarchical links between genetic and phenotypic changes, as the logic behind this tinkering.

The process whereby a gene or a genetic network can be recruited to perform new functions is usually ascribed to cooption. However, if developmental networks were dedicated mechanisms for making structures this explanation may be skin deep, implying, on the one hand, an ancestral role of each network that in the absence of reliable deep phylogenies would be difficult to ascertain and, on the other hand, some purposeful process or predetermined idea as to the novel structure that should be made by cooption (Minelli 2000). To say, for instance, that a *Pax-6*-dependent genetic network is a blueprint for making the eye (Halder et al. 1995; Gehring 2002), but it can be coopted for making a nose, tells nothing of how an eye or a nose actually becomes assembled. Given the role of *Pax-6* in regulating cell proliferation and differentiation in such diverse structures as parts of the central nervous system, lens and nasal placodes and pancreatic cells (Grindley et al. 1995; Matsunaga et al. 2000; Murakami et al. 2001; Madhavan et al. 2006; Zhang et al. 2006), the key to understanding its function in building either eyes or noses would lie ultimately in the developmental context related, in turn, to the phylogenetic history.

In the light of these observations, we may conclude that developmental genetic networks are assembled by development itself and the differential activation of genes which is detected during ontogeny only reveals the contingent adaptation of each cell to its microenvironment (Minelli 2003). According to this view, the 'developmental programme' would be the result of an enormous number of coupled mechanisms and would be subjected to canalization owing to spatiotemporal and molecular constraints (Galis et al. 2001; Siegal and Bergman 2002; Brakefield 2006). As a consequence, the phenotypic effects of genetic or epigenetic mutations will be unpredictable, unless we know the properties of the field (Goodwin 1984) which are related to, but not coincident with, the expression patterns of the genes and include, particularly, the above-mentioned developmental constraints.

6.4 Can NeoDarwinism Provide a Theory of EvoDevo?

As discussed so far, homology cannot be discriminated from homoplasy on merely genetic grounds. This conclusion is implicit in the gene-centred approach, insofar as similarities of gene expressions found in similar structures are considered to be an argument in favour of homology, while dissimilarities (belonging to homology by definition) cannot disprove homology. Given the mismatch between genes and phenotypes and the paucity of a fossil record illustrating transitional forms, this asymmetry is due to the assumption that the sameness which persists through variation is ultimately the body part itself. This judgment, however, is based on preconceptions, such as the belonging of the part to the same body plan (Holland and Holland 1999), a morphological definition which is based itself on homology. Somewhat paradoxically, then, the neoDarwinian mechanistic approach involves circularity and an essentialistic view of form which may be reminiscent of an idealistic definition of homology (Owen 1843). This starting point becomes necessary owing to insufficient databases, but more importantly to theoretic inadequacy. As stressed by several authors, the neoDarwinian paradigm does not produce any viable theory of development since it deals with inheritance and variation, but it does not account convincingly for character formation (Bowler 1989). By assuming that novelty and order should emerge from the interaction between 'inner' chance (random mutation) and 'outer' necessity (natural selection) over long periods of time, development itself, as an ordered process, should be ultimately the result of selection. This, indeed, is the rationale behind Haeckel's law: as ontogeny depends on phylogeny in terms of causation, evolutionary novelties should arise by fortuitous addition and embryonic stages, except for adaptive (caenogenetic) modifications, should correspond to adult ancestral forms. This conclusion clearly shows that in an evolutionary context the emphasis on selection as the predominant external cause of order causes a conflation of ancestor and archetype (Goodwin 1984). As a consequence, investigating homologies in order to get insights into phylogeny comes to be coincident with analysing modifications of ideal characters and 'developmental programmes' inherited from some prototypical form. To escape circularity, ontogeny should be released from the overriding control of phylogeny, that is, homologies should be considered as autonomous expressions of form.

6.5 EvoDevo Beyond NeoDarwinism

For this purpose, several strategies have been suggested. One of them is to deal separately with discrete levels of homology focussing attention on those which address more directly the origin of characters, such as biological or generative homology (Butler and Saidel 2000). However, to parcel up homology into separate units is not an answer to the problem, since an appropriate theory would be required in order to integrate generative and historical aspects into a coherent definition. Another suggestion is to consider the molecular products of genetic networks to be themselves homologous (Gilbert and Bolker 2001). Yet, this so-called process homology would give little insight into the origin and relationships of biological forms if, as it usually happens, the same basic molecular tool kit were used by every developing animal in its own way (Bolker and Raff 1996). One could object that all developmental changes should be ascribed ultimately to genetic changes, since cell activities are all governed by genes. However, in the light of the previous observations, to establish a causal connection would be a formidable task, and moreover, the opposite relationship cannot be excluded from a logical point of view. The chain of causes and effects goes back to the dawn of life, but the latter does not seem to be necessarily coincident with the appearance of encoding, replicable molecules, such as those of nucleic acids. On the contrary, it seems equally plausible that life arose when primitive feedback metabolic circuits were organized under the influence of environmental factors and became associated with spontaneously assembled double-layered membranes and stabilizing molecules, perhaps inorganic in the beginning (Fry 2000; Wright 2000; Vermeji 2006). In a similar vein, it has been suggested that prior to the advent of finely tuned genetic circuits morphogenesis of ancestral multicellular organisms could have been governed largely by 'generic' physicochemical forces (Newman and Müller 2000; Newman 2005; Newman et al. 2006). In this hypothetical 'pre-Mendelian world' a single genotype could map onto many phenotypes, giving perhaps an account for the Cambrian explosion of body plans. Only later phenotypes became stabilized by genes and natural selection came on the scene, until the integration of genetic and epigenetic mechanisms under selective control led to the establishment of 'organizational homologies', fundamental units of form above the genetic level acting as 'formal attractors of design, around which more design is added' (Müller and Newman 1999; Müller 2003). By suggesting a relative independence of biological forms from gene activities at the simplest and highest levels of organization, we should be able to rescue homology from the neoDarwinian shoehorn discussed above.

In this scenario, the control over the phenotype would shift during evolution from generic to genetic/selective to modular/organizational. However, the concept itself of control over morphology may be misleading, if we agree to consider biological forms to be nothing but an expression of *all* qualities and interactions of their components at any evolutionary or developmental stage (see Sect. 6.3). Contrary to the hypothesis of a higher morphological plasticity in a pre-Mendelian world, it seems conceivable that physical forces acting together with elementary developmental mechanisms such as asymmetric cell division, biochemical oscillation and dichotomous differentiation could lead to a restricted number of possible morphologies, and larger, rather than smaller, genetic changes could be necessary to bring about phenotypic changes. This conclusion may be corroborated by the finding that in simple biological systems, such as different species of the colonial alga *Volvox*, different genes and divergent developmental pathways come to produce very similar shapes (Kirk and Kirk 2004; Hallmann 2006) In any case, a relationship between the hypothetical pre-Mendelian world and Cambrian radiation of body plans seems unlikely, as the antiquity of the genome of Animalia (Ohno 1996), the constant morphs of the enigmatic Ediacaran fossils (Malakhov 2004) and the possibility of deciphering Early Cambrian animals and, tentatively, pre-Cambrian and Cambrian fossil embryos by current criteria (Conway Morris 1998a, b; Dong et al. 2005; Gostling et al. 2007), suggest that modern animals arose when the integration between generic and genomic processes was not so different from the present one.

For all these reasons, the notion of a variegated pre-Mendelian world does not seem to provide solid grounds for considering embryogenesis to be nothing but a stage of a quasi-cyclical process of development of form driven by generic mechanisms acting together with elementary cell activities and Darwinian competition between cells and developmental modules consisting of characters which have 'a life of their own' (Minelli 2003). This hypothesis has been elaborated in order to give a truly mechanistic account of embryogenesis avoiding the concept of a developmental genetic programme as well as recapitulation and finalism. Function itself should be purged of any crumb of hidden finalism. For instance, according to Minelli in ancestral lophotrochozoans and ecdysozoans, respectively, the role of cilia and cuticles might have been primarily one of 'generic' stabilization of form. However, this suggestion would push the dawn of modern superclades back to a hypothetical pre-Mendelian world, which could be questionable, as discussed above. In a similar way, any direct parallelism between generic and embryonic processes could be misleading, at least because neither gametes nor eggs can be compared to 'generic' cells.

Despite their merits, in my opinion these models do not succeed in providing an adequate account of embryogenesis as a process. As a consequence, they can give only limited insight into the mechanistic roots of homology, as developmental modules appear to be self-contained morphological entities which organize biological forms from 'above', either following some rules of design (Müller and Newman 1999; Müller 2003), or interacting between each other by internal Darwinian competition (Minelli 2003). In order to get deeper insights into these interactions and organizing properties, we should include an approach from 'below' into this process by addressing the origin of modules. Leaving aside all speculations on hypothetical affinities with ancestral developmental mechanisms, this could be done in the context of the embryology of Animalia at its current level of integration between generic and genetic processes. In this context, starting usually from the fertilized egg, developmental modular entities arise from the segregation of increasingly specified morphogenetic fields with concomitant increase of information and interactivity. Thus, from a general point of view embryogenesis might be described as a process of modular expansion by which each module becomes relatively independent of, but more subtly interconnected with, all other modules. While the latter relationships may correspond to different levels of control from 'above' on morphology, such as those brought about by organizational homologies (Müller 2003) and evolutionary stable character configurations (Wagner and Schwenk 2000), an approach from 'below' should start theoretically from the egg. Since at present the computation of the egg is beyond our ability, an alternative approach may be designed by analysing the argument which says that belonging to 'relatively similar body plans' is a prerequisite for homology (Holland and Holland 1999). This argument meets with circularity insofar as cladistic systematics and the gene-centred view of development consider body plans to be aggregates of 'essential' characters, and homologies to be variants of these characters (see Sects. 6.3, 6.4). By contrast, from the perspective of EvoDevo, homologies should be viewed as constrained sets of different, but related morphologies, and body plans as the products of peculiar configurations and interactions of developmental modules which represent the precondition for homology. As already suggested by classical embryology, the establishment of such configurations in bilateral animals may be rooted approximately in stages around gastrulation, preceding the phylotypic point of the developmental hourglass (Duboule 1994; Raff 1996). Putting aside types of development which presumably are derived, such as those of yolky eggs, we could observe that in isolecithal eggs gastrulation starts in the vegetal hemisphere, and that ectoderm and endoderm derive, respectively, from the animal and the vegetal half, and the possible sources of mesoderm, albeit more variable, follow recognizable patterns: ectomesoderm is anteroventral or ventrolateral, prostomial (schizocoelic) mesendoderm is posterodorsal or posterolateral, often being associated subsequently with segmentation and terminal growth, and gastral (enterocoelic) mesendoderm lies in an intermediate position closer to the vegetal pole (Fig. 6.3). This pattern, then, appears to be constrained, presumably by some conserved molecular and structural architecture of the egg. Taking into account that the role of a gene or a signalling pathway depends on the time and place of its activation (see Sect. 6.3), the molecular data on the networks involved in axial specification and germ-layer differentiation should be placed in the framework of this shared embryonic morphology. When compared in different phyla, gene expressions and reciprocal spatial relationships of the ectodermal, endodermal and mesodermal territories during gastrulation and initial tissue differentiation may allow us to discover how different developmental trajectories and combinations of the same basic modules and signalling systems prelude the appearance of constrained morphologies. To reconstruct the polarity of evolutionary changes, different embryonic geographies and their associated genetic patterns should be plotted on independently elaborated phylogenetic trees, such as those produced by phylogenomics. Insights could also be derived from palaeontology, on the condition that the fossil record is interpreted not only in the light of existing morphologies, but also with reference to phenotypes which might be predicted as the outcome of alternative combinations of modules. Of course, this approach is not straightforward, since the fossil record is fragmentary and the molecular trees are far from being conclusive, sometimes being influenced by underlying phylogenetic preferences arising from morphology. Another challenge would be to account for the derived patterns of embryogenesis, which should be included in a comprehensive theory together with the presumably primitive ones. It should also be stressed that a strictly mechanistic account of the evolution of development may be impossible to provide, as dissociation, fusion, duplication, loss and divergent differentiation of modules could be the result of many different genetic and epigenetic processes. In spite of these problems, when compared in the developmental context described above, molecular spatiotemporal and morphological data could be reciprocally illuminating. Throwing stronger light on the centrality of inherency and induction to the evolutionary process, they may reveal at least some fundamental developmental drives (Arthur 2001) which prelude the appearance of body plans.

This integrating approach to EvoDevo, which is different from Haeckelian recapitulation, but reminiscent of the law of embryonic divergence (von Baer 1828), could help the theory of evolution to move from a gene-centred selection towards a generative paradigm of the origin of biological forms (Goodwin 1984).

In order to illustrate in concrete terms these alternative views, the following sections analyse briefly their bearings on the problem of the origin of vertebrates.

6.6 EvoDevo and the Traditional Taxonomy of Protochordates: Current Achievements and Historical Roots

At present, it is almost universally accepted that vertebrates originated from ancestral invertebrate chordates, which were similar to living protochordates (lancelets and tunicates). Significantly, this phylogeny is linked to the rise of Darwinism, as Darwin and Haeckel attached the greatest importance to the purported transition from invertebrates to vertebrates that was revealed by the pioneering embryological studies of Kowalevsky (1866, 1867). According to Darwin (1871), these discoveries about the embryology of ascidians and lancelets supported the possibility 'that in an extremely remote period a group of animals existed resembling in many respects the larvae of our present Ascidians, which diverged into two great branches—the one retrograding in development and producing the present class of Ascidians, the other rising to the crown and summit of the animal kingdom by giving birth to the Vertebrata'. According to Haeckel (1868), they pointed towards the existence of ancestral worms (Chordaea) from which the vertebrates arose initially as Acrania, still represented by the skull-less amphioxus, while tunicates branched off evolving in a retrograde direction. Haeckel elaborated on this scenario in his subsequent books and used the term 'Prochordonier' (protochordates) to indicate the ancestors of vertebrates in his phylogenetic tree illustrating the evolution of man (Haeckel 1874b; Fig. 6.1).

Haeckel's and Darwin's interpretations were still influenced by the tradition of idealist morphology, including conceptions such as recapitulation and a progressive, anthropocentric view of evolution (Ospovat 1981). As a legacy from this tradition, the parallelism Owen (1866) established between the lancelet and the archetype of the vertebrates contributed to viewing the lancelet as a transitional form assembling some features of invertebrates, but all 'essential' characters of vertebrates (Fig. 6.2).

The enduring influence of these speculations can be detected, among other things, in the so-called new head hypothesis (Gans and Northcutt 1983; Northcutt and Gans 1983), which has proved to be extremely important in propelling the current programme of EvoDevo research on protochordates (Zimmer 2000). According to this view, cephalochordates arose from larval tunicates by paedomorphosis, involving an increase in size and complexity of the nervous system and the appearance of myotomes and coelomic cavities for a more efficient locomotion. Subsequently, a number of interconnected changes brought about by the muscularization of the hypomere and the evolution of the neural crest/placodes allowed for a more active, predatory style of life and led to the dawn of the vertebrates. In particular, the neural crest/placodes came to be implicated in the most prominent novelty of vertebrates, a 'new head' that was added to a headless lancelet-like ancestor.

In this scenario, Haeckelian echoes are the archetypal status of the lancelet as a prevertebrate and the progressive evolution by gradual addition. At the present time, both the lancelet and the ascidian tadpole are commonly designated by words such as 'prototypical', 'archetypal' and 'model'. These definitions imply that all 'essential' features of the body plan of vertebrates should be present in simpler form in protochordates. Therefore, in agreement with the gene-centred view of EvoDevo

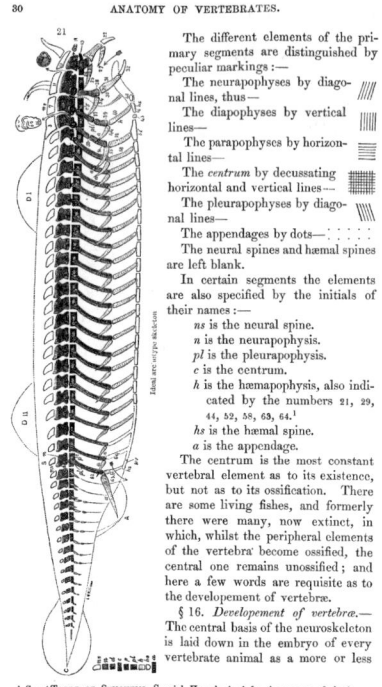

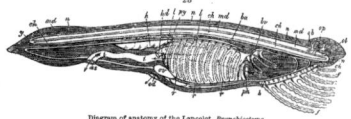

Fig. 6.2 The archetype of the vertebrates (*left*) according to Owen (1866). On the front page (*right*) an idealized scheme of the lancelet suggests a striking similarity to the archetype

the origin of vertebrates may be elucidated by comparing sequences and expression patterns of developmental genes. Besides developmental genetics, the idea of the archetype dominates comparative genomics as well as palaeontology. The genomic organization of the lancelet has been defined as 'archetypal' (Garcia-Fernàndez and Holland 1994), since it comprises single genes instead of multiple gene families, and particularly, a single Hox gene cluster which is similar to the Hox clusters of vertebrates. Assuming that increased complexity should be achieved by addition, the evolution of vertebrates has been related to two or more rounds of whole duplications of this prototypical genome and the subsequent cooption of genes allowing for the elaboration of ancestral characters and the evolution of innovations, such as the limbs and neural crest/placodes (Holland et al. 1994; Furlong and Holland 2002, 2004). Further evidence for this scenario should be provided by the fossil record, which in the past few years has yielded a good number of lancelet-like candidates for the ancestry of vertebrates (Shu et al. 1999, 2003; Shu 2003; Mallat and Chen 2003; Zhang and Hou 2004).

Another derivative of Haeckel's phylogeny (Fig. 6.1) is the definition of the lancelet as a 'transitional form between invertebrates and vertebrates'. Actually, a transition makes sense only if it is referred to stem groups. As the latter are still a matter of speculation, just as the phyletic lineages linking them to crown groups, the conception of a transitional form, as well as that of the lancelet as 'the closest invertebrate relative of vertebrates', is devoid of any factual meaning and could hardly be falsified.

The influence of these conceptions on current theories of the origin of vertebrates has not been modified significantly by structural studies casting doubt on several traditional homologies between cephalochordates and vertebrates (Gemballa et al. 2003; Ruppert 2005), nor by recently discovered genomic peculiarities, including the very large size and high posterior divergence of the Hox cluster, which have led to the lancelet being specified not as the ancestor, but as the best proxy to the ancestor of vertebrates (Ferrier et al. 2000; Minguillón et al. 2002, 2003, 2005; Jiménez-Delgado et al. 2006). Nor it has been modified significantly by molecular phylogenies, showing that tunicates may be more closely related than lancelets to vertebrates (Philippe et al. 2005; Delsuc et al. 2006; Vienne and Pontarotti 2006). Making comments on the latter findings, students of the lancelet have concluded that two model systems to study are better than one and that, in any case, the lancelet is the best available proxy of the basal chordates, as tunicates are derived from both the genomic and the morphological point of view (Schubert et al. 2006; Holland 2006; Holland and Holland 2007).

In the meanwhile, however, Northcutt (2005) has proposed a revised version of the new head hypothesis. Moving away from the idea of transformation of 'essential' characters, this novel hypothesis takes a step towards the view suggested in Sect. 6.5, since it focuses attention on the role of changes in embryonic tissue interactions in bringing about evolutionary changes. According to Northcutt, a crucial event which led to the dawn of the chordates was a redistribution of the germ layers during the formation of the blastula, in particular, a novel position of the mesoderm that came to be located closer to the animal pole than in more primitive fate maps, such as those of spiralians and ambulacrarian deuterostomes. Signalling from this dorsal mesoderm could trigger neuralization of the dorsal ectoderm, leading to the evolution of the neural plate. In turn, novel tissue interactions between the neural plate and adjacent non-neural ectoderm led to the appearance of the neural crest and neurogenic placodes. Unfortunately, this hypothesis is flawed by a number of shortcomings. Northcutt based his reasoning on compared fate maps of protostomes and deuterostomes at a generic blastula stage. However, this stage is variable in time even within the same species, and moreover, it does not allow one to predict the reciprocal positions of the germ layers which are achieved during gastrulation, when neural induction and other crucial tissue interactions occur. Moreover, the statement of Northcutt that in spiralians, just as in echinoderms, prospective mesoderm lies closer to the vegetal pole than in the chordates is not correct. As representatives of spiralians (protostomes) Northcutt quoted two fate maps of polychaetes taken from the treatise on embryology of Anderson (1973). However, in this book and in all other textbooks of comparative embryology, the position of the 4d blastomere progenitor of mesendoderm in spiralians is specified as 'posterodorsal', just as that of the prospective dorsal mesoderm of vertebrates and, indeed, it does not seem to be so closer to the vegetal pole than that of the mesoderm in the chordates, in any case, not closer than in protochordates (Fig. 6.3). Yet, Northcutt considers the

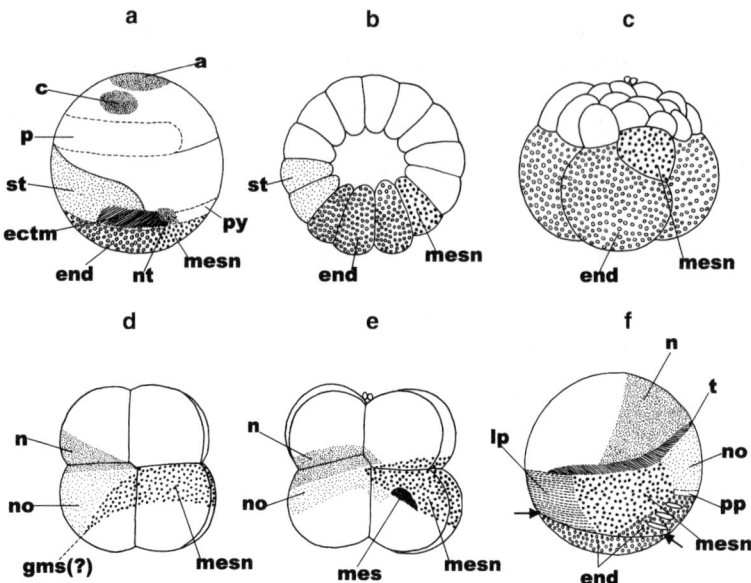

Fig. 6.3 Presumptive maps of two polychaetes (**a** *Podarke*, **b** *Eupomatus*), a gastropod (**c** *Crepidula*), a lancelet (**d** *Branchiostoma belcheri*), an ascidian (**e** *Halocynthia*) and an amphibian (**f** *Ambystoma*). Animal pole at the *top, anterior left*. *a* apical organ, *c* cerebral ganglion, *ectm* ectomesoderm, *end* endoderm, *gms* gastral mesoderm, *lp* lateral plate (ventral mesoderm), *mes* mesenchyme, *mesn* muscle (prostomial) mesendoderm, *n* nervous system, *no* notochord, *nt* neurotroch (posttrochal nervous system), *p* prototroch, *pp* mesoderm of the prechordal plate, *py* pygidium, *st* stomodaeum, *t* mesoderm of the tail. The *arrows* indicate the border of the future blastopore. Until now, the localization of gastral (enterocoelic) mesendoderm in the fate map of the lancelet has not been specified. In all these maps the prospective territories appear to be distributed according to a conserved pattern, in particular muscle mesendoderm is always found in posterodorsal position. However, the future notochord of lancelets and tunicates is anterior, just as the prospective stomodaeum-ectomesoderm of spiralians, while the notochord of vertebrates is found in posterior position. In a similar way, the future nervous system is anterior in protochordates, but posterior in vertebrates, corresponding to the posterodorsal ectoderm of spiralians. Together with that of another polychaete, *Scoloplos*, the fate map of *Podarke* (**a**) has been quoted by Northcutt (2005) in support of his thesis that in protostomes prospective mesoderm is closer to the vegetal pole than in the chordates. This general conclusion is not corroborated by the fate maps (compare **b**, **c** with **d**, **e**) (**a**, **b** After Anderson 1973; **c** after Conklin 1897; **d** after Tung et al. 1962; **e** after Nishida 1987; **f** after Balinsky 1965)

repatterning of the blastula bringing about novel tissue interactions to be essentially the same in protochordates and vertebrates. If so, it remains to be explained why the neural crest/placodes evolved in vertebrates, but not in lancelets and tunicates. Northcutt suggests that these innovations appeared as tissue primordia in basal chordates, but they could evolve into definite neural crest and neurogenic placodes 'after chordates passed a critical threshold in size'. This rather vague explanation does not produce insights from a mechanistic point of view, and moreover, it conflicts with the fossil record, as putative ancestral vertebrates do not appear to be larger than living protochordates (Shu et al. 1999, 2003; Shu 2003; Mallat and Chen 2003; Zhang and Hou 2004).

6.7 An Alternative View on the Phylogeny of Protochordates and the Origin of Vertebrates

The conclusions of the revised new head hypothesis are in agreement with the traditional phylogeny and the related idea that ancestral vertebrates were fishlike animals similar to the lancelet. However, a comparative analysis of the positions and gene expression patterns of the prospective territories in deuterostomes and gastroneuralians that was not limited to the fate maps, but that included the stages of gastrulation and neurulation, has produced a quite different picture (Raineri 1998, 2006). When the prospective maps are oriented with reference to the axis of the egg and the future body axes, protochordates appear to be different from vertebrates, but similar to gastroneuralians. As shown by the localization of the prospective territories and the expression of genes involved in dorsoventral specification, in lancelets and tunicates the notochord and nervous system are ventral. In agreement with this position, the mesoderm is shifted towards the vegetal side during gastrulation and the nervous system develops parallel to the closing blastopore, while in vertebrates the mesoderm is shifted towards the animal side and the nervous system forms apart from the blastopore (Figs. 6.3 and 6.4). Taking into account the molecular, embryological and structural evidence, I have concluded that lancelets and tunicates belong to

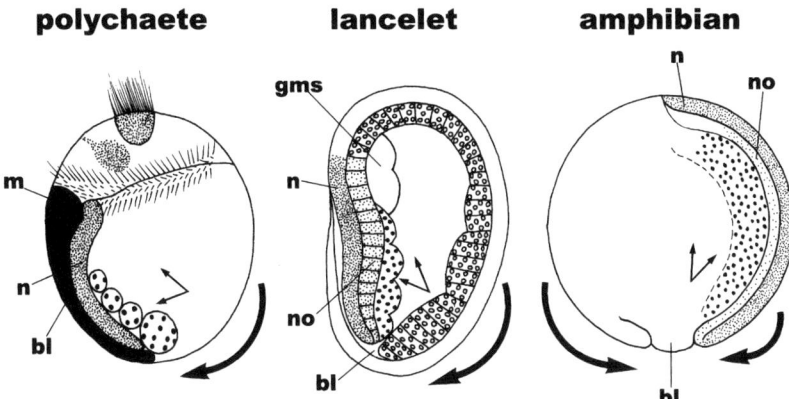

Fig. 6.4 Morphogenetic movements during gastrulation (endoderm not shown) in a polychaete (after Dawydoff 1928), a lancelet (after Cerfontaine 1906) and an amphibian (after Balinsky 1965). *bl* blastopore, *gms* gastral mesoderm, *m* mouth, *n* nervous system, *no* notochord. In the polychaete, just as in the lancelet, the more pronounced growth of the posterodorsal side shifts the blastopore from vegetal to ventral side and muscle mesoderm (*large dots*) towards the ventral side. The nervous system differentiates parallel to the closing blastopore from posterior to anterior in the polychaete, from anterior to posterior in the lancelet. At the end of gastrulation the residual opening of the blastopore forms the mouth in the polychaete and the neuroenteric canal in the lancelet. By contrast, in amphibians the growth of the posterodorsal lip of the blastopore towards the vegetal pole is less pronounced than that of the anteroventral lip, which moves over the yolky endoderm concomitantly with the ingression of the lateral plate and the expansion of ventral ectoderm. The dorsal mesoderm (prospective muscle and notochord) is shifted towards the animal side and the nervous system develops apart from the blastopore

two hitherto unrecognized phyla of gastroneuralians with no particular phylogenetic affinity to vertebrates. Just as in the hypothesis of Northcutt (2005), the great innovation of the vertebrates is suggested to be the central nervous system which evolved in relation to the novel dorsal position of the mesoderm, allowing, in turn, for the appearance of the neural crest/placodes. However, in contrast with the hypothesis of Northcutt, the latter are considered to be a derivative of the same primitive tissue that gave origin to the nervous systems of gastroneuralians, including lancelets and tunicates, and the dorsal position of mesoderm in vertebrates is ascribed primarily to a different pattern of gastrulation. These conclusions have profound phylogenetic implications. Contrary to the neoDarwinian paradigm, in this scenario ancestral vertebrates did not evolve from an archetypal chordate body plan by means of gradual genetic and phenotypic changes, but they appeared when conserved genetic networks came to be activated in a different spatiotemporal context that was created by embryonic modifications. This evolutionary process could lead to relatively rapid phenotypic changes, and moreover, it would show a high degree of freedom, as there would be no direct causal relationship between triggering factors and phenotypic outcomes of the embryonic modifications. However, being constrained by the properties of the evolving system itself, the results of this process might be partly predicted, as discussed earlier (see Sect. 6.5). For instance, taking into account the role of mesodermal signalling in neural induction and the peculiar ability of dorsal ectoderm to undergo invagination/expansion depending presumably on its position, the evolution of a nervous system like that of vertebrates could have been expected from a type of development shifting mesoderm towards the dorsal side (Raineri 2006). This predictability, and the underlying logic, is linked to the origin of homology, which is to be sought in similar combinations of modules, not necessarily in similar forms. This means that stem groups are not archetypes of the crown groups, an obvious conclusion that in the case of vertebrates has been ignored owing to the overwhelming influence of the concept of the chordates. If this conception is devoid of real foundations, the evolutionary history of vertebrates could be very different from the story that has been told until now. Rather than being fishlike animals with a tiny head, stem vertebrates could have had, for instance, a bipartite body plan including 'head' and 'tail', similar to that of calcichordates or vetulicolians (Jefferies et al. 1996; Aldridge et al. 2007), a nervous system consisting of ganglia besides the incipient dorsal nerve cord, and a hydraulic axial support that only later evolved into solid notochord.

References

Albrecht-Buehler G (1990) In defense of "nonmolecular" cell biology. Int Rev Cytol 120:191–241
Aldridge RJ, Xian-Guang H, Siveter DJ, Siveter DJ, Gabbott SE (2007) The systematics and phylogenetic relationships of vetulicolians. Palaeontology 50:131–168
Anderson DT (1973) Embryology and phylogeny in annelids and arthropods. Pergamon, Oxford
Arthur W (1997) The origin of animal body plans: a study in evolutionary developmental biology. Cambridge University Press, Cambridge

Arthur W (2001) Developmental drive: an important determinant of the direction of phenotypic evolution. Evol Dev 3:263–270
Ashe HL, Levine M (1999) Local inhibition and long-range enhancement of Dpp signal transduction by Sog. Nature 398:427–431
Balinsky BI (1965) An introduction to embryology. Saunders, Philadelphia
Ben-Tabou de-Leon S, Davidson EH (2006) Deciphering the underlying mechanism of specification and differentiation: the sea urchin gene regulatory network. Sci STKE. doi:10.1126/stke.3612006pe47
Bolker JA, Raff RA (1996) Developmental genetics and traditional homology. BioEssays 18: 489–494
Bowler PJ (1989) Evolution: the history of an idea. University of California Press, Berkeley
Brakefield PM (2006) Evo-devo and constraints on selection. Trends Ecol Evol 21:362–368
Butler AB, Saidel WM (2000) Defining sameness: historical, biological, and generative homology. BioEssays 22:846–853
Cerfontaine P (1906) Recherches sur le développement de l'*Amphioxus*. Arch Biol 22:229–418
Coffman JA, McCarthy JJ, Dickey-Sims C, Robertson AJ (2004) Oral-aboral axis specification in the sea urchin embryo: II. Mitochondrial distribution and redox state contribute to establishing polarity in *Strongylocentrotus purpuratus*. Dev Biol 273:160–171
Conklin EG (1897) The embryology of *Crepidula*. J Morphol 13:3–209
Conway Morris S (1998a) The crucible of creation: the Burgess Shale and the rise of animals. Oxford University Press, Oxford
Conway Morris S (1998b) Eggs and embryos from the Cambrian. BioEssays 20:676–682
Conway Morris S (2000) Evolution: bringing molecules into the fold. Cell 100:1–11
Conway Morris S (2003) The Cambrian "explosion" of metazoans and molecular biology: would Darwin be satisfied? Int J Dev Biol 47:505–515
Darwin C (1871) The descent of man, and selection in relation to sex. Murray, London
Davidson EH, Erwin DH (2006) Gene regulatory networks and the evolution of animal body plans. Science 311:796–800
Davidson EH, Rast JP, Oliveri P, Rasnick A, Calestani C, Yuh C-H, Minokawa T, Amore G, Hinman V, Arenas-Mena C, Otim O, Brown CT, Livi CB, Lee PY, Revilla R, Rust AG, Pan Z, Schilstra MJ, Clarke PJC, Arnone MI, Rowen L, Cameron RA, McClay DR, Hood L, Bolouri H (2002) A genomic regulatory network for development. Science 295:1669–1678
Dawydoff C (1928) Traité d'embryologie comparée des invertébrés. Masson, Paris
De Robertis EM, Sasai Y (1996) A common plan for dorsoventral patterning in Bilateria. Nature 380:37–40
Delsuc F, Brinkmann H, Chourrout D, Philippe H (2006) Tunicates and not cephalochordates are the closest living relatives of vertebrates. Nature 439:965–968
Dong X, Donoghue PCJ, Cunningham JA, Liu J, Cheng H (2005) The anatomy, affinity, and phylogenetic significance of *Markuelia*. Evol Dev 7:468–482
Duboule D (1994) Temporal colinearity and the phylotypic progression: a basis for the stability of a vertebrate Bauplan and the evolution of morphologies through heterochrony. In: Akam M, Holland P, Ingham P, Wray G (eds) The evolution of developmental mechanisms. Company of Biologists, Cambridge, pp 135–142
Erwin DH, Davidson EH (2002) The last common bilaterian ancestor. Development 129: 3021–3032
Ferrier DEK, Minguillón C, Holland PWH, Garcia-Fernàndez J (2000) The amphioxus Hox cluster: deuterostome posterior flexibility and *Hox14*. Evol Dev 2:284–293
Fry I (2000) The emergence of life on Earth: a historical and scientific overview. Rutgers University Press, New Brunswick
Furlong RF, Holland PWH (2002) Were vertebrates octoploid? Philos Trans R Soc Lond Ser B 357:531–544
Furlong RF, Holland PWH (2004) Polyploidy in vertebrate ancestry: Ohno and beyond. Biol J Linn Soc 82:425–430
Galis F, Sinervo B (2002) Divergence and convergence in early embryonic stages of metazoans. Contrib Zool 71:101–113

Galis F, van Alphen JJM, Metz JAJ (2001) Why five fingers? Evolutionary constraints on digit number. Trends Ecol Ev 16:637–646

Gans C, Northcutt RG (1983) Neural crest and the origin of vertebrates: a new head. Science 220:268–274

Garcia-Fernàndez J, Holland PWH (1994) Archetypal organization of the amphioxus Hox gene cluster. Nature 370:563–566

Gehring WJ (2002) The genetic control of eye development and its implications for the evolution of the various eye-types. Int J Dev Biol 46:65–73

Gemballa S, Weitbrecht GW, Saanchez-Villagra MR (2003) The myosepta in *Branchiostoma lanceolatum* (Cephalochordata): 3D reconstruction and microanatomy. Zoomorphology 122:169–179

Gilbert SF, Bolker JA (2001) Homologies of process and modular elements of embryonic construction. J Exp Zool B Mol Dev Evol 291:1–12

Gilbert SF, Opitz JM, Raff RA (1996) Resynthesizing evolutionary and developmental biology. Dev Biol 173:357–372

Goodwin BC (1984) Changing from an evolutionary to a generative paradigm in biology. In: Pollard JW (ed) Evolutionary theory: paths into the future. Wiley, Chichester, pp 99–120

Gostling NJ, Donoghue PCJ, Bengtson S (2007) The earliest fossil embryos begin to mature. Evol Dev 9:206–207

Gould SJ (1977) Ontogeny and phylogeny. Harvard University Press, Cambridge

Grindley JC, Davidson DR, Hill RE (1995) The role of *Pax-6* in eye and nasal development. Development 121:1433–1442

Haeckel E (1866) Generelle Morphologie der Organismen. Reimer, Berlin

Haeckel E (1868) Natürliche Schöpfungsgeschichte. Reimer, Berlin

Haeckel E (1874a) The gastraea-theory, the phylogenetic classification of the animal kingdom and the homology of the germ-lamellae. Q J Microsc Sci 14:142–165

Haeckel E (1874b) Anthropogenie oder Entwickelungsgeschichte des Menschen. Englemann, Leipzig

Halder G, Callaerts P, Gehring WJ (1995) New perspectives on eye evolution. Curr Opin Gen Dev 5:602–609

Hallmann A (2006) Morphogenesis in the family Volvocaceae: different tactics for turning an embryo right-side out. Protist 157:445–461

Holland LZ, Holland ND (2007) A revised fate map for amphioxus and the evolution of axial patterning in chordates. Integr Comp Biol 47:360–372

Holland ND, Holland LZ (1999) Amphioxus and the utility of molecular genetic data for hypothesizing body part homologies between distantly related animals. Am Zool 39:630–640

Holland P (2006) My sister is a sea squirt? Heredity 96:424–425

Holland PWH, Garcia-Fernàndez J, Williams NA, Sidow A (1994) Gene duplications and the origins of vertebrate development. In: Akam M, Holland P, Ingham P, Wray G (eds) The evolution of developmental mechanisms. Company of Biologists, Cambridge, pp 125–133

Jacob F (1977) Evolution and tinkering. Science 196:1161–1166

Jefferies RPS, Brown NA, Daley PEJ (1996) The early phylogeny of chordates and echinoderms and the origin of chordate left-right asymmetry and bilateral symmetry. Acta Zool 77:101–122

Jiménez-Delgado S, Crespo M, Permanyer J, Garcia-Fernàndez J, Manzanares M (2006) Evolutionary genomics of the recently duplicated amphioxus Hairy genes. Int J Biol Sci 2:66–72

Kirk MM, Kirk DL (2004) Exploring germ-soma differentiation in *Volvox*. J Biosci 29:143–152

Kowalevsky A (1866) Entwickelungsgeschichte der einfachen Ascidien. Mem Acad Imp Sci St Petersbourg VII Ser 10:1–19

Kowalevsky A (1867) Entwickelungsgeschichte des *Amphioxus lanceolatus*. Mem Acad Imp Sci St Petersbourg VII Ser 11:1–17

Love AC, Raff RA (2006) Larval ectoderm, organizational homology, and the origins of evolutionary novelty. J Exp Zool B Mol Dev Evol 306:18–34

Madhavan M, Haynes TL, Frisch NC, Call MK, Minich CM, Tsonis PA, Del Rio-Tsonis K (2006) The role of Pax-6 in lens regeneration. Proc Natl Acad Sci USA 103:14848–14853

Malakhov VV (2004) New ideas on the origin of bilateral animals. Russ J Mar Biol 30(Suppl 1): 22–33

Mallat J, Chen J (2003) Fossil sister group of craniates: predicted and found. J Morphol 258:1–31

Matsunaga E, Araki I, Nakamura H (2000) *Pax6* defines the di-mesencephalic boundary by repressing *En1* and *Pax2*. Development 127:2357–2365

Matus DQ, Pang K, Marlow H, Dunn CW, Thomsen GH, Martindale MQ (2006) Molecular evidence for deep evolutionary roots of bilaterality in animal development. Proc Natl Acad Sci USA 103:11195–11200

Matus DQ, Pang K, Daly M, Martindale MQ (2007) Expression of Pax gene family members in the anthozoan cnidarian, *Nematostella vectensis*. Evol Dev 9:25–38

McGhee JD (2000) Homologous tails? Or tales of homology? BioEssays 22:781–785

Meaburn KJ, Misteli T (2007) Chromosome territories. Nature 445:379–381

Minelli A (1998) Molecules, developmental modules, and phenotypes: a combinatorial approach to homology. Mol Phylogen Evol 9:340–347

Minelli A (2000) Limbs and tail as evolutionarily diverging duplicates of the main body axis. Evol Dev 2:157–165

Minelli A (2003) The development of animal form: ontogeny, morphology, and evolution. Cambridge University Press, Cambridge

Minguillón C, Ferrier DEK, Cebrián C, Garcia-Fernàndez J (2002) Gene duplication in the prototypical cephalochordate amphioxus. Gene 287:121–128

Minguillón C, Jiménez-Delgado S, Panopoulou G, Garcia-Fernàndez J (2003) The amphioxus Hairy family: differential fate after duplication. Development 130:5903–5914

Minguillón C, Gardenyes J, Serra E, Filipe L, Castro C, Hill-Force A, Holland PWH, Amemiya CT, Garcia-Fernàndez J (2005) No more than 14: the end of the amphioxus Hox cluster. Int J Biol Sci 1:19–23

Müller GB (2003) Homology: the evolution of morphological organization. In: Müller GB, Newman SA (eds) Origination of organismal form. Beyond the gene in developmental and evolutionary biology. MIT Press, Cambridge, pp 51–69

Müller GB, Newman SA (1999) Generation, integration, autonomy: three steps in the evolution of homology. In: Homology, Novartis Foundation Symposium 222. Wiley, Chichester, pp 65–79

Müller GB, Wagner GP (1996) Homology, *Hox* genes, and developmental integration. Am Zool 36:4–13

Murakami Y, Ogasawara M, Sugahara F, Hirano S, Satoh N, Kuratani S (2001) Identification and expression of the lamprey *Pax6* gene: evolutionary origin of the segmented brain of vertebrates. Development 128:3521–3531

Newman SA (2005) The pre-Mendelian, pre-Darwinian world: Shifting relations between genetic and epigenetic mechanisms in early multicellular evolution. J Biosci 30:75–85

Newman SA, Müller GB (2000) Epigenetic mechanisms of character origination. J Exp Zool B Mol Dev Evol 288:304–317

Newman SA, Forgacs G, Müller GB (2006) Before programs: the physical origination of multicellular forms. Int J Dev Biol 50:289–299

Nishida H (1987) Cell lineage analysis in ascidian embryos by intracellular injection of a tracer enzyme. III. Up to the tissue restricted stage. Dev Biol 121:526–541

Northcutt RG (2005) The new head hypothesis revisited. J Exp Zool B Mol Dev Evol 304:274–297

Northcutt RG, Gans C (1983) The genesis of neural crest and epidermal placodes: a reinterpretation of vertebrate origins. Q Rev Biol 58:1–28

Oda-Ishii I, Bertrand V, Matsuo I, Lemaire P, Saiga H (2005) Making very similar embryos with divergent genomes: conservation of regulatory mechanisms of *Otx* between the ascidians *Halocynthia roretzi* and *Ciona intestinalis*. Development 132:1663–1674

Ohno S (1996) The notion of the Cambrian pananimalia genome. Proc Natl Acad Sci USA 93:8475–8478

Ospovat D (1981) The development of Darwin's theory. Natural history, natural theology, and natural selection, 1838–1859. Cambridge University Press, Cambridge

Owen R (1843) Lectures on the comparative anatomy and physiology of the invertebrate animals. Longman, Brown, Green and Longmans, London

Owen R (1866) The anatomy of vertebrates, vol 1. Longmans, Green, London
Philippe H, Lartillot N, Brinkmann H (2005) Multigene analyses of bilaterian animals corroborate the monophyly of Ecdysozoa, Lophotrochozoa, and Protostomia. Mol Biol Evol 22:1246–1253
Raff RA (1996) The shape of life. Genes, development, and the evolution of animal form. University of Chicago Press, Chicago
Raineri M (1998) Proposta di una nuova classificazione di Tunicati e Cefalocordati come Gastroneuralia. Implicazioni filogenetiche e cenni storici sulle origini del concetto di Protocordati. Ann Mus Civ St Nat G Doria 92:1–83
Raineri M (2006) Are protochordates chordates? Biol J Linn Soc 87:261–284
Reed RD, Chen P-H, Nijhout FH (2007) Cryptic variation in butterfly eyespot development: the importance of sample size in gene expression studies. Evol Dev 9:2–9
Richardson MK, Hanken J, Gooneratne ML, Pieau C, Raynaud A, Selwood L, Wright GM (1997) There is no highly conserved embryonic stage in the vertebrates: implications for current theories of evolution and development. Anat Embryol 196:91–106
Roux W (1895) Gesammelte Abhandlungen über Entwickelungsmechanik der Organismen. Engelmann, Leipzig
Ruppert EE (2005) Key characters uniting hemichordates and chordates: homologies or homoplasies? Can J Zool 83:8–23
Saint-Hilaire ÉG (1818) Philosophie anatomique. Méquignon-Marvis, Paris
Schierenberg E (2001) Three sons of fortune: early embryogenesis, evolution and ecology of nematodes. BioEssays 23:841–847
Schubert M, Escriva H, Xavier-Neto J, Laudet V (2006) Amphioxus and tunicates as evolutionary model systems. Trends Ecol Evol 21:269–277
Shu D-G (2003) A palaeontological perspective of vertebrate origin. Chin Sci Bull 48:725–735
Shu D-G, Conway Morris S, Zhang X-L, Hu S-X, Chen L, Han J, Zhu M, Li Y, Chen L-Z (1999) Lower Cambrian vertebrates from south China. Nature 402:42–46
Shu D-G, Conway Morris S, Han J, Zhang Z-F, Yasul K, Janvier P, Chen L, Zhang X-L, Liu J-N, Li Y, Liu H-Q (2003) Head and backbone of the Early Cambrian vertebrate *Haikouichthys*. Nature 421:526–529
Siegal ML, Bergman A (2002) Waddington's canalization revisited: developmental stability and evolution. Proc Natl Acad Sci USA 99:10528–10532
Spemann H (1938) Embryonic development and induction. Yale University Press, New Haven
Stern CD (2006) Evolution of the mechanisms that establish the embryonic axes. Curr Opin Genet Dev 16:413–418
Tung TC, Wu SC, Tung YYF (1962) The presumptive areas of the egg of amphioxus. Sci Sin 11:629–644
Vermeji GJ (2006) Historical contingency and the purported uniqueness of evolutionary innovations. Proc Natl Acad Sci USA 103:1804–1809
Vienne A, Pontarotti P (2006) Metaphylogeny of 82 gene families sheds a new light on chordate evolution. Int J Biol Sci 2:32–37
von Baer KE (1828) Entwickelungsgeschichte der Tiere. Borntrager, Königsberg
von Kölliker A (1864) Über die Darwin'sche Schöpfungstheorie. Zeit Wiss Zool 14:174–186
Wagner GP (1996) Homologues, natural kinds and the evolution of modularity. Am Zool 36:36–43
Wagner GP, Schwenk K (2000) Evolutionarily stable configurations: functional integration and the evolution of phenotypic stability. In: Hecht MK et al (eds) Evolutionary biology, vol 31. Kluwer/Plenum, New York, pp 155–217
Wright BE (2000) A biochemical mechanism for nonrandom mutations and evolution. J Bacteriol 182:2993–3001
Zhang X, Rowan S, Yue Y, Heaney S, Pan Y, Brendolan A, Selleri L, Maas RL (2006) *Pax6* is regulated by Meis and Pbx homoproteins during pancreatic development. Dev Biol 300:348–357
Zhang X-G, Hou X-G (2004) Evidence for a single median fin-fold and tail in the Lower Cambrian vertebrate, *Haikouichthys ercaicunensis*. Evol Biol 17:1162–1166
Zimmer C (2000) In search of vertebrate origins: beyond brain and bone. Science 287:1576–1579

Part III
Knowledge

Chapter 7
Overturning the Prejudices about Hydra and Metazoan Evolution

Hiroshi Shimizu

Abstract *Hydra*, a member of the phylum Cnidaria, has been used as a model organism for zoological research since the days of Trembley (*Mémoires pour Servir à l'Histoire d'un Genre de Polypes d'Eau Douce*, 1744). Anatomical observations and their theoretical interpretation provided common knowledge about *Hydra*: (1) diffusion plays a major role in the formation of a positional information gradient; (2) diffusion plays a dominant role in circulatory and digestive processes in *Hydra*; (3) the diffuse nerve net is a remnant of the ancient and functionally meager nervous system; (4) the digestive tract of hydra is a closed sac with only one opening working both as the mouth and as the anus. In this chapter, I describe our experimental results in recent years that led us to conclude that those widely accepted views are more or less erroneous.

7.1 Diffusion: Potentially an Ideal Mechanism for Material Transport in Primitive Metazoans

7.1.1 A Theoretical Consideration

Hydra is a member of the class Hydrozoa of the phylum Cnidaria. Hydrozoa is considered the class which appeared relatively recently in comparison with other classes, e.g., Anthozoa, Scyphozoa and Cubozoa (Fig. 7.1). *Hydra* inhabits freshwater, e.g., inland lakes and ponds, and feeds on plankton. The body of *Hydra* is made up of a long and cylindrical body column with a head at one end and a foot at the other end (Fig. 7.2a). Since the foot end, termed the "basal disc," is a sheet of epithelium that adheres to the substrate such as leaves, the body plan of hydra is

H. Shimizu
National Institute of Genetics, 1111 Yata, Mishima, 411-8540 Shizuoka, Japan
hshimizu@lab.nig.ac.jp

Fig. 7.1 Phylogenetic tree of the phylum Cnidaria

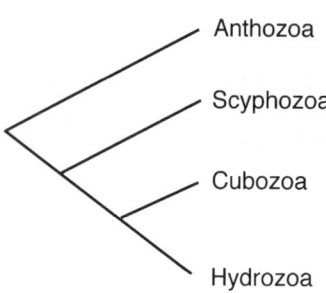

thought to be a closed sac with an opening at the head end termed the "hypostome." When Hydrozoa was considered the most ancient class (Haeckel 1874), the simple external morphology of hydra was thought to reflect the fact that evolution occurred from a structurally simple organism (hydra) to those with a complicated structure (sea anemone of class Anthozoa). This view has been proved incorrect by several means, including molecular phylogeny (Bridge et al. 1995).

Hydra's strong regenerative capacity has initiated pattern formation studies. Experimental and theoretical analyses carried out to elucidate the basic mechanism of regeneration resulted in the formulation of a positional information gradient in hydra that informs cells of their position in the animal (Wolpert 1969). The theory was later given a mathematical framework by Gierer and Meinhardt (1972) on the basis of diffusion-induced instability proposed by Turing (1952). In both concepts, transport of substances by diffusion was the key mechanism for the generation of the gradient. This was understandable since in a system of less than 1 cm diffusion is appropriate for molecules with a molecular weight of hundreds to 1,000 (Crick 1970). Besides these undertakings, diffusion was also thought to be a key mechanism in the physiological processes of hydra. In the processes of digestion, circulation, excretion, aspiration, etc., the gastric cavity of hydra provides the space where diffusion of digestive enzymes, digested materials, excreted materials, oxygen, etc. could occur. If diffusion is powerful enough, hydra may be able to live without any specific organs, e.g., the heart, small intestine, kidney, lung, etc., and related activities. Why did researchers put such a heavy load on diffusion? There is a possible reason. Since diffusion itself is a physical phenomenon, diffusible transport does not consume ATP. Such a simple but ultimately cost-efficient mechanism, therefore, is supposed to be a very useful mechanism for primitive organisms that have no distinct organs to carry out physiological functions. As a result, there has been no extensive research on the physiological processes of hydra most likely because the research might result in studying the mechanism of diffusion in living organisms, which sounds like a modest but not so attractive research theme.

7.1.2 Problems of the "Diffusion Paradigm"

Despite its tremendous presence, there have been no direct attempts to visually observe diffusion, implying that diffusion still has no credit. A simple way to observe

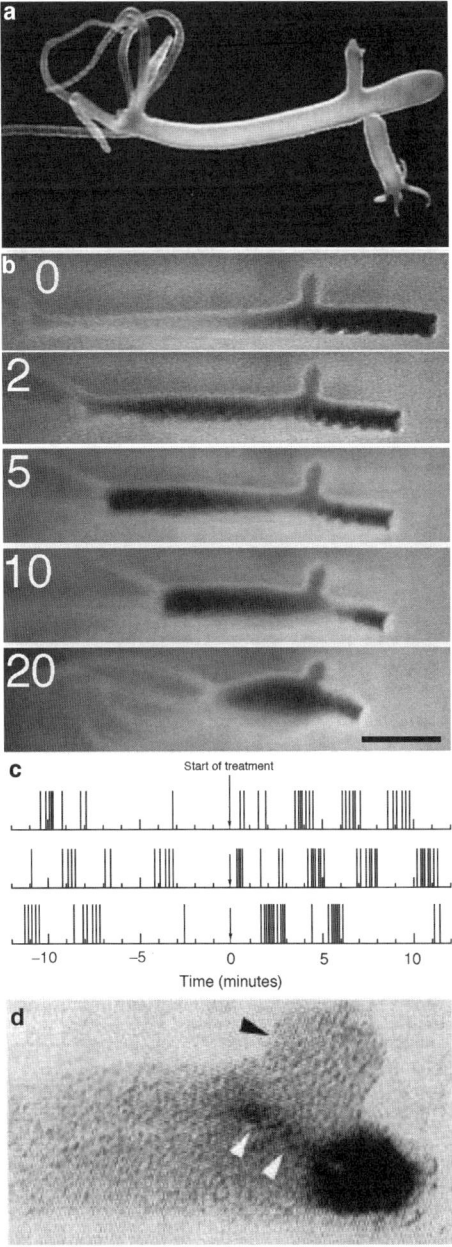

Fig. 7.2 a Polyp of *Hydra magnipapillata* (strain 105). **b** Change of distribution of the darkly stained gastrovascular fluid either in the peduncle region or in the whole body column. *Numbers* represent relative time in seconds. **c** Change of frequency of contraction by treating animals with KPHLRGRFamide solution (1×10^{-6} M). The pattern of contraction during 12 min before and 12 min after the start of treatment is shown for three cases. **d** Expression of *CnNk-2* in the peduncle region of hydra viewed by whole-mount in situ hybridization. The *black arrowhead* represents the bud and *white arrowheads* represent the expression at the base of the bud that later develops into the peduncle but that already expresses *CnNk-2*. *Bars* represent 1 mm

diffusion would be to put a small volume of dye or India ink into the system to see how it can spread; however, even such attempts have not been undertaken. Also, it remains unknown where diffusion could occur; intracellular space, interstitial space, gastric cavity or elsewhere. More seriously, a lot of cnidarians have a much bigger body size than hydra (jellyfish and sea anemones reach 30 cm in size). In those animals, diffusion-based transport is obviously insufficient for long-range internal transport. Nevertheless, no practical alternative mechanism has been proposed for such big cnidarians. In conclusion, it is a mystery for the author why researchers did not care about this issue while caring a lot about gene products, e.g., secreted proteins that can also diffuse in the organism.

7.1.3 Is Diffusion Powerful Enough for Circulation in Hydra?

In vertebrates, transportation of oxygen and nutrients throughout an animal is carried out by the cardiovascular system. Hydra apparently do not need such a system because diffusion of nutrients in the short digestive tract is considered modest but powerful enough, whereas oxygen can be supplied from the surrounding medium by diffusion across the surface of the epithelium. These considerations have supported the view that no specific circulatory organs or related activities are needed in hydra.

In "coelenterate" hydra, "coelenteron" (hollow cavity) is also called the "gastrovascular cavity," implying that the cavity serves as the space for circulation and digestion. To visualize the process of diffusion thereby outlining the cavity, we injected a small volume of India ink into the cavity of an elongate hydra polyp using a thin glass tube (Shimizu and Fujisawa 2003). We anticipated that the whole cavity would be darkly labeled. Surprisingly for the author, the labeling was not spread all over the animal but was restricted to the peduncle region of the elongate animal (Fig. 7.2b). Considering that the body column of hydra is potentially a tube, we next examined why there is labeling only in the peduncle but not in the main body column. A longitudinal section of the animal showed that the epithelial tissue is very thin in the peduncle, while it is relatively much thicker in the main body column, allowing fluid to stay only in the peduncle region of the cylindrical column. Hydra contracts occasionally, changing its shape from a long and thin cylinder to a fat and short bell shape. By the contraction, the main body column region of the cavity widened and the whole gastrovascular cavity was labeled by the India ink (Fig. 7.2b). This demonstrates that the gastrovascular cavity is in a "coelenteron"-like status only when the animal contracts and that the fluid in the peduncle can circulate throughout the cavity not by diffusion but possibly by a turbulent flow of fluid.

The contractile movement of the animal that changed the pattern of the distribution of the fluid was found to have several things in common with the pumping of the heart of higher organisms. First, treatment of hydra with KPHLRGRFamide (Mitgutsch et al. 1999), a neuropeptide originally derived from hydra, elevated the frequency of contraction (Fig. 7.2c). RFamide-type neuropeptides were originally purified and identified in clams by Price and Greenberg (1977) as cardioactive peptides. When cardiac tissue was treated with the peptides, the pumping frequency of the molluscan heart was elevated. Thus, pumping movements in molluscan heart and hydra peduncle were both elevated by RFamide-type neuropeptides. The second is the involvement of *CnNk*-2. *Nkx*-2.5 (Lints et al. 1993) and its orthologues are expressed in cardiac mesoderm and pharyngeal endoderm, both of which show pumping behavior (Chen and Fishman 2000). The tissue that expresses *Nkx*-2.5 orthologues pumps blood through the cardiovascular space in vertebrates and flies or pumps ingested food through the digestive tract in nematodes (Avery and Horvitz 1989; Okkema et al. 1997). In hydra, *CnNk*-2 is expressed in the peduncle (Grens et al. 1996), where pumping movement is observed extensively (Fig. 7.2d). In total, both peduncle of hydra and metazoan heart show pumping movement that occurs by the action of cardioactive neuropeptides, while expressing *Nkx*-2.5 orthologues. Therefore, unlike previously thought, hydra undergoes circulation of gastric fluid not by diffusion but by a pumping movement of the animal as in higher metazoans. We hypothesize that the blood vascular system connected to the cardiac system is a later invention after atmospheric oxygen concentration rose to a certain extent as described below.

When the evolutionary origin of the heart is discussed, the heart is naturally defined as an organ that pumps blood through the vascular system. This reflects the view that in animals that have no blood there is no organ which is functionally equivalent to the heart. The present observation provides an example that circulation of fluid even in the absence of a blood vascular system occurs by a mechanism that involves a pumping movement and factors related to heart pumping. We propose that the blood vascular system was of limited importance early in the Proterozoic Era, when the atmospheric oxygen concentration was significantly lower than the present level and blood-mediated oxygen transport was too costly. The cardiac system which developed in that era was possibly involved in nutrient transport and not in oxygen transport. With the rise of the oxygen level in the late Proterozoic Era (Canfield and Teske 1996), the blood vascular system appeared and merged with the preexisting cardiac system to become the cardiovascular system, thus transporting both nutrients and oxygen. Hydra could thus represent a rare example where the cardiac system existed separately from the blood vascular system.

Then a question arises as to whether the pumping movement occurs in Anthozoa, which is ancestral to Hydrozoa. In *Nematostella vectensis*, a burrowing sea anemone (Robson 1957; Fig. 7.3a), there is a movement that occurs specifically in the aboral region. *Nematostella* is a sediment dweller and it buries the aboral half of the body column into the sediment, while it captures prey with the tentacles in the oral end (Hand and Uhlinger 1994). For the purpose of dwelling in the sediment, it digs a hole into the sediment by a peristaltic movement (Trueman and Ansel 1969). During the process a constriction appears in the mid body column and this contractile ring is transferred into aboral orientation, showing a traveling wavelike appearance (Fig. 7.3b). By this movement, hydropressure is supplied to the tissue between the aboral end and the site of constriction. This hydropressure is considered absolutely necessary for pushing sediment material aside the animal while it is digging the hole. In that sense, the constriction-induced hardening of the aboral part of the polyp can be considered as a typical example of a hydrostatic skeleton, also termed "hydrostat"

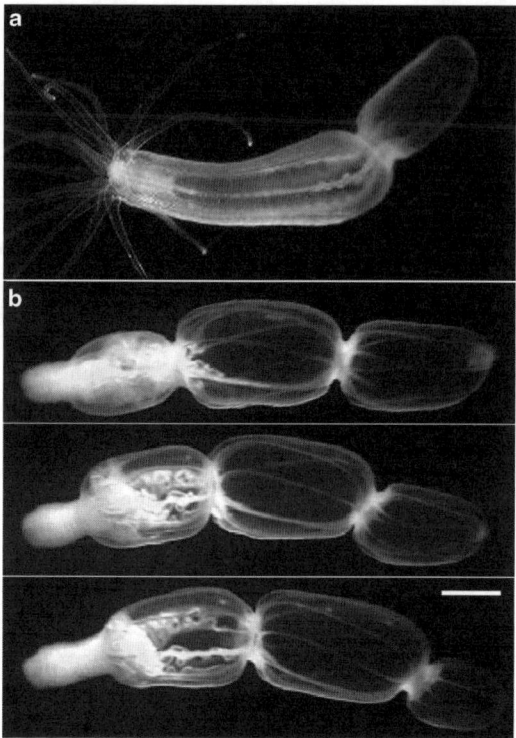

Fig. 7.3 a *Nematostella vectensis*. **b** Burrowing peristalsis in *Nematostella vectensis*. Images recorded every 30 s are shown. The *bar* represents 5 mm

(Smith and Kier 1989), and the pumping activity is exhibited not as a circulatory tool as in the heart of mammals and hydra, not as a pumping tool to send bolus into the digestive tract as in the pharynx of nematodes, but as a pressurizing pump to provide hardening of tissue as the hydrostat. Thus, the functional significance of pumping is very diverse among organisms.

7.1.4 Is Diffusion Powerful enough for Digestion in Hydra?

In organisms bigger in size, e.g., vertebrates, the digestive tract is basically a long tube. There the transport of bolus, digested materials and feces through the tract by peristalsis is a crucial issue in the digestive function because inactive transport results in capitulation, which is the most popular symptom in the digestive organs of *Homo sapiens*. Material transport by means of peristalsis is therefore generally thought to have become well organized as the digestive tract got longer. In hydra it is common knowledge that digestive enzymes, digested materials and waste materials can all travel through the short digestive tract by diffusion, implying that there is no need for peristalsis.

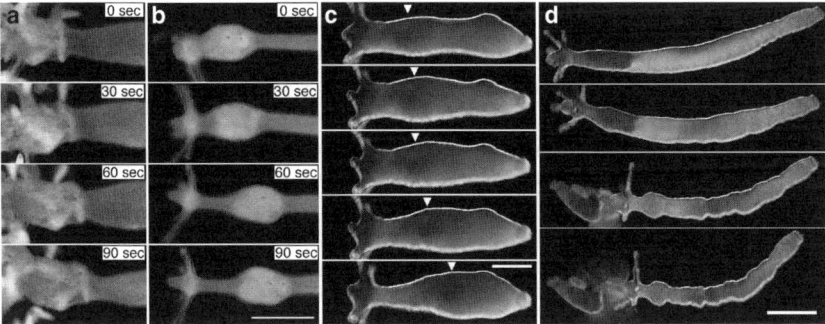

Fig. 7.4 a Intact hydra polyp fully fed with *Artemia*. **b** Esophageal reflex movement that occurs in polyps fed with five *Artemia*. For **a** and **b**, the *uppermost part* shows the polyp 3 min after feeding, and subsequent images show changes recorded 30, 60 and 90 s later. **c** Peristaltic reflex movement recorded every 10 s. **d** Defecation reflex movement. Images were recorded 0, 5, 20 and 30 s after the onset of defecation. *Bars* represent 1 mm

On observing how the ingested food is transferred in the cavity, mingled with digestive enzymes, and how waste is ejected from the hypostome, we became convinced that diffusion is never enough to result in digestive function in hydra (Shimizu et al. 2004). Hydra polyps when fed with *Artemia* expand the gastrovascular cavity to a great extent (Fig. 7.4a). It is generally accepted that continuous swallowing of and stuffing the cavity with *Artemia* is the driving force for the transfer of ingested *Artemia* forward. We found by feeding hydra polyps with a limited number of *Artemia* that the transfer of the ingested *Artemia* occurs in oral to aboral orientation without stuffing and pushing them forward from the back (Fig. 7.4b). This transfer apparently occurs by the extension of the body column just ahead of the ingested *Artemia* and simultaneously the narrowing of the column just behind the *Artemia*. This movement is similar to the movement of the mammalian esophagus that occurs on sending bolus from the pharynx to the stomach, which is known as "esophageal reflex" (Roesler 1960). The movement was therefore also termed "esophageal reflex movement." Movement that occurred in the next few hours transferred the contents back and forth in the cavity and was similar in appearance to the movement of intestinal tissue of mammals (Cannon 1912; Hukuhara et al. 1958, 1961a, b), and hence was termed "peristaltic reflex movement" (Fig. 7.4c) although the contents moved back and forth in the cavity because of the closed sac type structure. By this movement, fibrous material was condensed and formed feces. The feces was finally ejected from the mouth as the result of movement termed "defecation reflex movement" (Fig. 7.4d). This movement was similar in morphology to the movement of the mammalian rectum (Takaki et al. 1987). It is technically possible to reduce the peristaltic movement by tissue manipulation (Shimizu, unpublished results). There the formation of solid feces did not occur and the excrement was similar in appearance to diarrhea, demonstrating that the peristaltic reflex movement plays an essential role in making solid feces in hydra. It may be worthwhile investigating the underlying mechanism of these movements. In hydra it is possible to construct polyps having no neurons and to maintain them by feeding by hand (Marcum and Campbell 1978). Since such "nerve-free hydra" showed only a weak peristaltic movement, we conclude that the movements are mostly neurological events as is the case with mammals, where absence of enteric neurons provokes the absence of peristalsis (Swenson 2002). It is important to note here that the direction of transfer in the three movements is not the same in hydra, which is in contrast to mammals, where all the movements occur in one direction.

7.2 Does the "Diffuse Nerve Net" Have a Function in Hydra?

7.2.1 Common Knowledge

Neurologically, there is common knowledge about hydra. The nervous system of hydra is known as the "diffuse nerve net." This net is made up of single neural cell bodies scattered in the body column which are connected to each other by single neural fibers (not plural fibers). This nerve net has no apparent polarity since an external stimulus applied provokes contraction all over the animal and this contraction does not depend upon the site of the stimulus. This implies that the neurons of the nerve net conduct the stimulus in all directions. To add to that, single nerve cells are symmetrical in shape, with typically two neural fibers extending from a cell body, thus showing no sign of morphological polarity (Hager and David 1997). These observations led to the view that the diffuse nerve net represents the remnant of the most ancient nervous system (Matthews 1997).

7.2.2 Problems Emerging

Although the nervous system of higher organisms is more organized, structurally complicated and more abundant in neurons, structural simplicity and functional meagerness are not directly linked to each other. For example, the reticular formation located at the dorsal side of the human brainstem is made up of a diffuse nerve net (Hobson and Drazier 1980). This system is known to play crucial roles in aspiration, circulation and other activities, although the actual mechanism of how the net-like structure is related to regulatory function remains unknown. It is therefore inappropriate to conclude that fewer neurons are indicative of fewer functions.

7.2.3 Diffuse Nerve Net as an Enteric Nervous System

The present observation (Fig. 7.4) clearly demonstrates that the diffuse nerve net functions as an enteric nervous system (equivalent to Auerbach's plexus of the

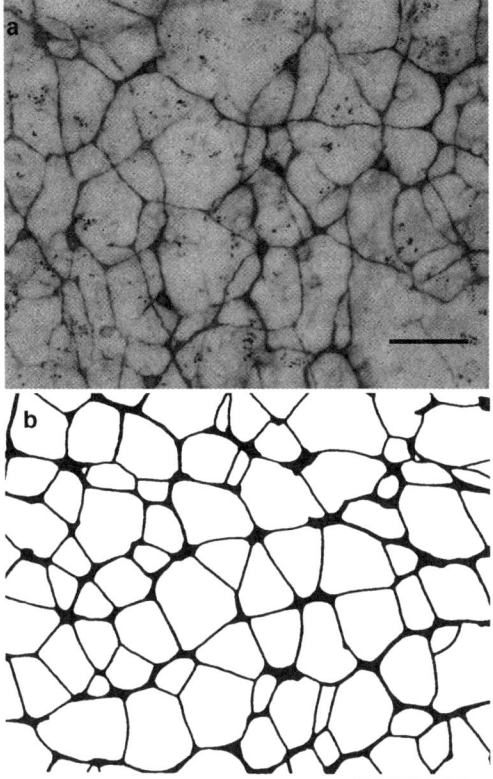

Fig. 7.5 a *Hydra* diffuse nerve net in the body column. **b** Enteric nervous system (Auerbach's plexus) of the guinea pig rectum. *Bars* represent 50 μm (**a**) and 2 mm (**b**), respectively

mammalian digestive tract) that regulates digestive movements (Gabella 1979). Structurally, the diffuse nerve net of hydra consists of single neurons and single neural fibers (Sakaguchi et al. 1996; Fig. 7.5a), which is in contrast to the mammalian enteric nervous system made up of ganglions and bundles of neural fibers connecting them (Fig. 7.5b). Although the enteric nervous system of hydra is meager in the number of neurons, we do not consider this meagerness to be a problem. The intestinal tissue of the mammalian intestine is made up of dozens of layers of longitudinally and circumferentially running smooth-muscle fibers (Gershon and Erde 1981). Hydra body column tissue is also made up of longitudinally running muscle fibers in the ectoderm and circumferentially running fibers in the endoderm (Mueller 1950). However, the fibers are both one-layered. We hypothesize that hydra's nerve net consisting of single neurons is related to one-layered muscle fibers and therefore is not related to poor functions. It is important to note that the digestive movements in the mammalian digestive tract are one-directional and each of the neurons has polarized morphology unlike the bipolar morphology of neurons of hydra.

A question emerges here as to why in the primitive phylum having a nervous system such a functionally mature nervous system is already present. One possibility we suggest is that ancient classes, e.g., class Anthozoa of Cnidaria, have a more primitive system. Indeed, sea anemones do not undergo digestive movements and seem to have no nervous system functionally equivalent to the enteric nervous system.[1] In the body column of anthozoan polyps there is a diffuse nerve net (Hirose, unpublished results) which possibly plays a role in the burrowing peristalsis already mentioned in Sect. 7.1.3 but that is obviously not as sophisticated as in hydra, which shows at least three types of movement. An alternative possibility could be that there is an extinct phylum that had a more primitive nervous system.

It may be worthwhile to mention the significance of enteric nervous system in digestion in other organisms. Nematodes have no enteric nervous system in the intestine. The pharynx of nematodes undergoes an autonomous pumping movement that is related to *ceh*-22 (ref), an *Nkx*-2.5 orthologue, and this movement provides physical force which is strong enough to send ingested food through the digestive tract. The peristaltic movement that can be seen in the intestine of nematodes is thought to be a passive movement evoked by the pumping movement of the pharynx. The absence of the enteric nervous system in nematodes could thus represent the result of degeneration of the system. The same situation is also found in tunicates which belong to Urochordata.

7.3 Body Plan of Hydra: Closed Sac or a Tube as in Higher Metazoans?

7.3.1 Common Knowledge

A common understanding of how the tubelike digestive tract appeared in evolution is that it evolved from the closed sac type digestive tract found in cnidarian polyps. Hydra ingests food from the hypostome and ejects feces also from the hypostome. The hypostome therefore plays two roles, both as mouth and as anus, although the

---

[1] In Anthozoa, which is ancestral to Hydrozoa, polyps that have a tubular digestive tract are not found. Polyps of corals and sea anemone undergo digestion by septal filament tissue located at the edge of the septum. There, prey is engulfed from the mouth live and is captured by nematocytes that are distributed on the surface of cnidoglandular tract at the edge of the septal filaments. The prey captured and immobilized is sandwiched by cnidoglandular tissue and flagellated tissue and later digested in the tiny space between the two tissues. During the process, apparently no peristalsis-like movement is seen to occur. The nervous system distributed there does not form a net-like structure (Shinzato 2007; Hirose, unpublished results). Therefore, the feature of digestion in anthozoan polyps is quite different from the process in Hydrozoa. This is supportive for a possibility that the digestive process in hydra is not orthologous to the digestive process in vertebrates despite their amazing structural and behavioral similarities but represents an example of convergent evolution. Ach, 5-HT and other known factors that evoke digestive movements in the mammalian digestive tract (Price et al. 1969; Hiatt et al. 1970) have no effect in hydra (Fujisawa, personal communication) and this may also reflect the evolutionary relationship between them.

hypostomal opening is called the mouth. Haeckel (1874) considered this property as evidence that the common ancestor was an organism that had a closed sac or gastrula-like digestive tract.

7.3.2 Problems Emerging

From the viewpoint of developmental phylogeny, it may not be appropriate to call this sole opening the mouth. When one takes a look at the embryogenesis of *Hydractinia echinatta*, another member of the class Hydrozoa, the embryo passes through a planula stage before becoming a polyp. The planula larva swims primarily in one direction and under the effect of certain types of bacterial products and subsequent signaling by neurotransmitters, the planula attaches its swimming front end to the substrate, typically the surface of shells of hermit crabs (Spindler and Müller 1972). The planula then undergoes metamorphosis to form a polyp. When this occurs, the head, including the hypostome, is formed at the end opposite the site of attachment, namely, the swimming rear end. If one defines the anterior–posterior polarity, one definition could be that the posterior end is the supposed rear end of an animal. Should the opening formed at the rear end nonetheless be termed the mouth and not the anus?

Similar problems have also been raised from molecular analysis. Around the anterior and oral end of developing embryos or adults of higher organisms, there is a similar pattern of expression of non-HOX-type homeobox gene *Nkx-2.5* or its orthologues at the anterior end, whereas there is a pattern of expression of *Wnt-3a* or its orthologues at the posterior end (Fig. 7.6). *Wnt-3a* is known as the gene expressed in the blastopore and hence is considered as a posterior marker (Klecker and Niehrs 2001). In hydra, the hypostomal tip expresses the *Wnt-3a* orthologue (Hobmayer et al. 2000). So far, hydra is the only exception where the *Wnt-3a* orthologue is expressed at the so-called oral end.

Anatomically, Kanajew (1928) reported the presence of a potential opening at the aboral end of hydra, demonstrating that the end of the sac is not closed. This opening, termed "porus aboralis" or "aboral pore," however, was not regarded with popularity, but was occasionally neglected. Even in the acclaimed zoology textbook by Hyman (1940), hydra is treated as an animal having a closed sac system and there is no reference to the pore. Interestingly, she referred to the aboral pore of anthozoan polyps on several occasions in the book. In addition, in the phylum Ctenophora, which is ancestral to the phylum Cnidaria, an opening at the aboral end is immediately noticed. This opening, termed "anal pore," is the site where a fraction of the digested material is ejected (Ruppert and Barnes 1996). Therefore, it is quite often the case that there is a potential opening at the aboral end of Cnidaria and Ctenophora. Nevertheless, hydra was specialized as an animal having a closed sac body plan as a representative of the phylum Cnidaria.

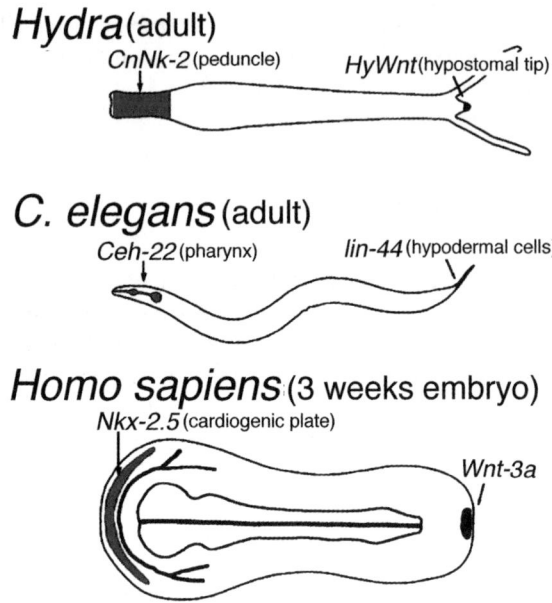

Fig. 7.6 Pattern of expression of *Nkx*-2.5 and *Wnt*-3a homologues in adult *Hydra*, adult *Caenorhabditis elegans* and mouse embryo at stage nine

7.3.3 Experimental Analysis

Observation of the basal disc of hydra in tissue sections (Shimizu et al. 2007) showed that there is an area of tissue at the center of the disc that has no extracellular matrix (ECM) (Fig. 7.7a). The immunohistochemical analysis had shown that the ECM intervenes two epithelia of hydra, namely, the ectoderm and the endoderm (Shimizu et al. 2002). However, we did not notice the hole in the ECM around the aboral pore because whole-mount samples were always laid down horizontally, making it difficult to see the pore. To examine how this situation with no ECM is maintained, we carried out analysis at various levels. Our concern was that the aboral pore is formed where bud detachment occurs in the final stage of asexual reproduction and this might suggest that the aboral pore formation is the artifact accompanied by budding. To test this possibility we examined polyps propagated by sexual reproduction. Although observing the embryogenesis of hydra is technically difficult because the whole embryo is tightly covered by a cuticular structure, newly hatched polyps examined showed that the pore is present (Fig. 7.7b), thus suggesting that the aboral pore represents the innate morphology of the animal. Another issue to be examined was how the aboral pore once formed is maintained. Hydra matrix metalloprotease (HMMP) has an activity to degrade the ECM and its activity is particularly high in the tentacles and the foot region of hydra (Leontovich et al. 2000). To test if the hole of the mesoglea is maintained by the degradation of mesoglea by this HMMP activity, we measured the HMMP activity around the

7 Overturning the Prejudices about Hydra and Metazoan Evolution 129

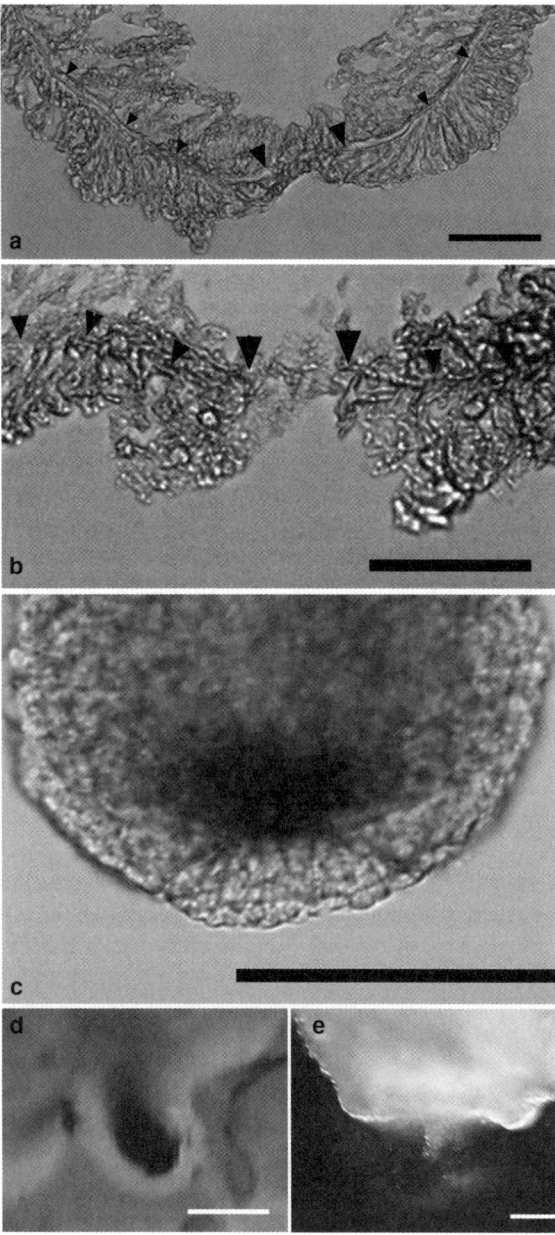

Fig. 7.7 a Histology section of the center of the basal disc that shows the area of no intervening of the extracellular matrix (ECM). **b** Histology section in polyps that underwent sexual reproduction. In **a** and **b**, *large arrowheads* represent the edge of the ECM and *small arrowheads* represent the position of the ECM. **c** Expression of hydra matrix metalloprotease messenger RNA at the basal disc region revealed by whole-mount in situ hybridization. **d** Ingestion of external fluid (stained black by India ink) through the pore observed when India ink solution was dropped onto the basal disc of a floating polyp of strain N10. **e** Ejection of digested material through the aboral pore. The *bars* in **a**, **d** and **e** represent 0.5, 1.0 and 1.0 mm, respectively

aboral pore and compared it with the activity in other parts of the animal. It was shown that the activity is conspicuously high around the aboral pore compared with other regions, e.g., tentacles (Fig. 7.7c), thus suggesting that the "ozone hole" like absence of the ECM is maintained by the degradation of the ECM by HMMP.

Despite the absence of the ECM, observation under a dissecting microscope suggested no opening of the pore. To determine whether there is transfer of materials, we carefully investigated the behavior of hydra. As a result, both ingestion and ejection of material were found to occur occasionally, although at a low frequency and in a strain-specific manner. Ingestion occurred in two situations. First, contraction of the polyp evoked the ingestion. In one situation, ingestion occurred in polyps floating on the surface of the culture solution. Polyps move around on the bottom and along the sidewall by a somersaulting motion. As a result they occasionally reach the surface of the culture solution and attach themselves to the surface by the basal disc and keep floating in an upside-down status. A small volume of India ink was dropped onto the surface of the polyps. More occasionally in strains N10 and N11, which have a bigger fraction of the basal disc than wild-type strains, the ink went directly into the gastrovascular cavity, thereby darkly staining the whole cavity (Fig. 7.7d).

Ejection was monitored several hours after feeding because digested material was easily discernable as a maker for gastric contents. When a fed polyp attached to the bottom of the dish was carefully detached, we found that digested material is ejected through the aboral pore (Fig. 7.7e). When the polyp in this situation was fixed and basal disc tissue was excised and examined under the microscope, the pore was seen to be wide open, demonstrating that detachment did not injure the tissue to provoke the release of material through the injury. These observations demonstrate that the aboral pore has no function as the anus, but that material transfer in both directions occurs.

Although functionally meager, we propose that this aboral pore and oral opening of other metazoans share a common ancestral origin because of the similarities listed below:

1. Both structures have direct connection to the digestive tract.
2. Both structures are located at one end of the animal, whose adjacent tissue (peduncle of hydra and pharynx of other metazoans) expresses *Nkx-2.5* orthologues either in the embryonic stage or in the adult stage.
3. The tissue that expresses *Nkx-2.5* orthologues shows pumping activity.

These observations suggest a possibility that the digestive tract of Cnidaria was originally a tube as in other metazoans. For a sessile life style as in hydra, attaching to the substrate is a critical issue together with feeding. When both were required, a possible strategy that was taken might have been to assign the attaching function to the "anterior" end and also to assign feeding and excreting function to the "posterior" end. Jellyfish and sessile sea anemones apparently have a closed sac (the burrowing sea anemone *Nematostella vectensis* has an aboral pore (Hyman 1940)). This could be a result of adaptation to a sessile and drifting life style. Self-feeding planuloids seem to have a closed sac digestive tract. From the view of planuloid theory (the common ancestor was of planuloid type), the closed sac system might have appeared in a planuloid-type ancestor and in polyp-type animals the tube type digestive tract was already present.

7.4 Discussion

Hydra's simple body plan and its position in the evolutionary tree produced a lot of common but unproved knowledge. In the author's view, this lack of proof is not because it was technically difficult to prove, but because scientists had been trapped by the prejudice that lower invertebrates are furnished with minimal physiological activities and related organs therefore depend upon diffusion. Hence, there were no attempts to examine the diffusion paradigm. We examined those views by very simple behavioral and visual analyses and reached the current conclusion that hydra does not rely on diffusion for circulation and digestion but it does rely upon similar kinetic mechanisms to those used in vertebrates. The digestive and circulatory behaviors of hydra had characteristics common to those of mammals, showing similar movements, involving orthologous genes, etc. This suggests that metazoan evolution was not the process from the system with fewer physiological organs and related mechanisms to the system with more of everything. A similar tendency is found on genomic levels. It seems that scientists had been affected by the prejudice that primitive organisms have a smaller genome size and a lower number of genes and introns. The result of the genome project for *Nematostella vectensis* overturned these prejudices drastically (Putnam et al. 2007). As mentioned briefly, the number of genes, arrangement of gene clusters, number of introns of *Nematostella vectensis*, etc. were all relatively similar to those of the human genome more than any other genome ever examined. The total number of protein-coding genes of *Nematostella vectensis* is estimated to be about 18,000, compared with 22,000 in the human genome and 14,000 in *Drosophila* and *Caenorhabditis elegans*. Therefore, the smaller genome size and fewer introns we find in invertebrates turned out not to represent the intermediate stage of genome size increase from lower invertebrates to mammals as previously thought but to be the result of loss of genes and introns somewhere during metazoan evolution from Urmetazoa to mammals.

It was known for many years that the genome size of hydra is bigger than that of other invertebrates is and relatively closer to that of human (Zacharias et al. 2004). However, this size was considered to reflect the fact that hydra undergoes asexual budding that can give rise to a high rate of somatic mutations, implying that the hydra genome is abundant in pseudogenes that emerged by somatic mutations. When the hydra genome project was being planned, there was a discussion about which strain should be used in the project (H. Bode and R. Steele, personal communication). Because of its exceptionally and relatively small genome size compared with that of *Hydra magnipapillata* (Zacharias et al. 2004), *Hydra viridissima* was also a candidate strain simply because it was thought that the large size of the *Hydra magnipapillata* genome was of little significance.

Taken together, it seems reasonable to assume that hydra has both a genomic and a physiological background that is not so far distant from that of mammals. Are the genomic and physiological characteristics inherited from a common ancestor? If so, we will need to think of a novel scenario of metazoan evolution. For example, the common ancestor had all the basic genetic components (hardware) and the mechanism that links the components to physiological functions (software). It is likely that the mechanism developed as adaptation to the environment of the earth, but the basis has not changed much. Therefore, we need to investigate how the mechanism developed and became complicated on the basis of similar genomic architecture.

References

Avery L, Horvitz HR (1989) Pharyngeal pumping continues after laser killing of the pharyngeal nervous system of Caenorhabditis elegans. Neuron 3:473–485

Bridge D, Cunningham CW, DeSalle R, Buss LW (1995) Class-level relationships in the phylum Cnidaria: Molecular and morphological evidence. Mol Biol Evol 12:679–689

Canfield DE, Teske A (1996) Late Proterozoic rise in atmospheric oxygen concentration inferred from phylogenetic and sulphur-isotope studies. Nature 382:127–132

Cannon WB (1912) Peristalsis, segmentation and the myenteric reflex. Am J Physiol 30:114–128

Chen JN, Fishman MC (2000) Genetics of heart development. Trends Genet 16:383–388

Crick F (1970) Diffusion in embryogenesis. Nature 231:420–422

Gabella G (1979) Innervation of the gastrointestinal tract. Int Rev Cytol 59:129–193

Gershon MD, Erde SM (1981) The nervous system of the gut. Gastroenterology 80:1571–1594

Gierer A, Meinhardt H (1972) A theory of biological pattern formation. Kybernetik 12:30–39

Grens A, Gee L, Fisher DA, Bode HR (1996) CnNK-2, an NK-2 homeobox gene, has a role in patterning the basal end of the axis in hydra. Dev Biol 180:473–488

Haeckel E (1874) Die Gastrea Theorie, die phylogenetische Classification des Tierreiches und die Homologie der Keimblätter. Jena Z Naturwiss 8:1–55

Hager G, David CN (1995) Pattern of differentiated nerve cells in hydra is determined by precursor migration. Development 124:569–576

Hand C, Uhlinger K (1994) The unique, widely distributed sea anemone, Nematostella vectensis Stephenson: a review, new facts, and questions. Estuaries 17:501–508

Hiatt RB, Goodman I, Overweg NI (1970) Serotonin and intestinal motility. Am J Surg 119:527–529

Hobmayer B, Rentzsch F, Kuhn K, Happel CM, von Laue CC, Snyder P, Rothbacher U, Holstein TW (2000) WNT signalling molecules act in axis formation in the diploblastic metazoan Hydra. Nature 407:186–189

Hobson JA, Brazier MAB (eds) (1980) The reticular formation revisited. Raven, New York, pp 55–66

Hukuhara T, Yamagami M, Nakayama S (1958) On the intestinal intrinsic reflexes. Jpn J Physiol 8:9–20

Hukuhara T, Sumi T, Kotani S (1961a) The role of the ganglion cells in the small intestine taken in the intestinal intrinsic reflex. Jpn J Physiol 11:281–288

Hukuhara T, Kotani S, Sato G (1961b) Effects of destruction of intramural ganglion cells on colonic motility: possible genesis of congenital megacolon. Jpn J Physiol 11:635–640

Hyman LH (1940) The invertebrates. Protozoa through Ctenophora. McGraw-Hill, New York, pp 662–695

Kanajew J (1928) Über den Porus aboralis bei Pelmatohydra oligactis Pall. Zool Anz 76:37–44

Klecker C, Niehrs C (2001) A morphogen gradient of Wnt/b-catenin signalling regulates antero-posterior neural patterning in Xenopus. Development 128:4189–4201

Leontovich AA, Zhang J, Shimokawa K, Nagase H, Sarras MP (2000) A novel hydra matrix metalloproteinase (HMMP) functions in extracellular matrix degradation, morphogenesis and the maintenance of differentiated cells in the foot process. Development 127:907–920

Lints TJ, Parsons LM, Hartley L, Lyons I, Harvey RP (1993) Nkx-2.5: a novel murine homeobox gene expressed in early heart progenitor cells and their myogenic descendants. Development 119:419–443

Marcum BA, Campbell RD (1978) Development of Hydra lacking nerve and interstitial cells. J Cell Sci 29:17–33

Matthews GG (1997) Neurobiology: molecules, cells, and systems. Blackwell, Cambridge, pp 23–24

Mitgutsch C, Hauser F, Grimmelikhuijzen CJ (1999) Expression and developmental regulation of the Hydra-RFamide and Hydra-L Wamide preprohormone genes in Hydra: evidence for transient phases of head formation. Dev Biol 207:189–203

Mueller JF (1950) Some observations on the structure of hydra, with particular reference to the muscular system. Trans Am Microscop Soc 69:133–147

Okkema PG, Ha E, Haun C, Chen W, Fire A (1997) The Caenorhabditis elegans NK-2 homeobox gene ceh-22 activates pharyngeal muscle gene expression in combination with pha-1 and is required for normal pharyngeal development. Development 124:3965–3973

Price WE, Shehadeh Z, Thompson GH, Underwood LD, Jacobson ED (1969) Effects of acetylcholine on intestinal blood flow and motility. Am J Physiol 216:343–347

Price DA, Greenberg MJ (1977) Purification and characterization of a cardioexcitatory neuropeptide from the central ganglia of a bivalve mollusc. Prep Biochem 7:261–281

Putnam NH, Srivastava M, Hellsten U, Dirks B, Chapman J, Salamov A, Terry A, Shapiro H, Lindquist E, Kapitonov VV, Jurka J, Genikhovich G, Grigoriev IV, Lucas SM, Steele RE, Finnerty JR, Technau U, Martindale MQ, Rokhsar DS (2007) Sea anemone genome reveals ancestral eumetazoan gene repertoire and genomic organization. Science 317:86–94

Robson E (1957) A sea-anemone from brackish water. Nature 179:787–788

Roesler H (1960) Esophageal reflex origin of myocardial infarction. Am J Med Sci 240:159–162

Ruppert EE, Barnes RD (1996) Invertebrate zoology. Saunders College, Fort Worth, pp 164–167

Sakaguchi M, Mizusina A, Kobayakawa Y (1996) Structure, development, and maintenance of the nerve net of the body column in Hydra. J Comp Neurol 373:41–54

Shimizu H, Fujisawa T (2003) Peduncle of Hydra and the heart of higher organisms share a common ancestral origin. Genesis 36:182–186

Shimizu H, Zhang X, Zhang J, Leontovich A, Fei K, Yan L, Sarras MP Jr (2002) Epithelial morphogenesis in hydra requires de novo expression of extracellular matrix components and matrix metalloproteinases. Development 129:1521–1532

Shimizu H, Koizumi O, Fujisawa T (2004) Three digestive movements in Hydra regulated by the diffuse nerve net in the body column. J Comp Physiol A 190: 623–630

Shimizu H, Takaku Y, Zhang X, Fujisawa T (2007) The aboral pore of hydra: evidence that the digestive tract of hydra is a tube not a sac. Dev Genes Evol 217:563–568

Shinzato C (2008) PhD thesis, James Cook University

Smith KK, Kier WM (1989) Trunks, tongues and tentacles: moving with skeletons of muscle. Am Sci 77:28–35

Spindler KD, Müller WA (1972) Induction of metamorphosis by bacteria and by lithium-pulse in the larvae of Hydractinia echinata (Hydrozoa). Wilhelm Rouxs Arch 169:271–280

Swenson O (2002) Hirschsprungs disease: a review. Pediatrics 109:914–918

Takaki M, Neya T, Nakayama S (1987) Functional role of lumbar sympathetic nerves and supraspinal mechanism in the defecation reflex of the cat. Acta Med Okayama 41:249–257

Trueman ER, Ansell AD (1969) The mechanisms of burrowing into soft substrata by marine animals. Oceanogr Mar Biol Annu Rev 7:315–366

Turing AM (1952) The chemical basis of morphogenesis. Philos Trans R Soc Lond Ser B 237:37–72
Wolpert L (1969) Positional information and the spatial pattern of cellular differentiation. J Theor Biol 25:1–47
Zacharias H, Anokhin B, Khalturin K, Bosch TC (2004) Genome sizes and chromosomes in the basal metazoan Hydra. Zoology (Jena) 107:219–227

Chapter 8
The Search for the Origin of Cnidarian Nematocysts in Dinoflagellates

Jung Shan Hwang, Satoshi Nagai, Shiho Hayakawa, Yasuharu Takaku, and Takashi Gojobori

Abstract The phylum Cnidaria is thought to be unique among animals as it contains a nematocyst or cnidocyst, which is a stinging organelle used for prey capture, defense and movement. Questions have been raised regarding the evolution of nematocysts since nematocyst-like structures can be found in some protists. In particular, the nematocyst of *Polykrikos kofoidii* or *Polykrikos schwartzii* structurally resembles the stenotele of *Hydra* at the electron microscope level. Both structures not only share morphological resemblance, but also the manner of catching prey. Evidence also suggests that a part of the *Polykrikos* nematocyst is assembled with a protein similar to *Hydra* minicollagen. In this chapter, we summarize our findings to date and discuss the evolutionary processes that may underlie the similarities of the nematocysts of *Polykrikos* and *Hydra*.

8.1 Background

The phylum Cnidaria includes a diverse group of animals with more than 10,000 species, including jellyfish, sea anemones, sea pens, hydroids, corals, etc. These animals have a simple diploblastic body plan, and generally consist of two cell layers (ectoderm and endoderm) and a central digestive cavity with an opening mouth surrounded by tentacles. Cnidaria is divided into four classes (Hydrozoa, Scyphozoa, Anthozoa and Cubozoa), which mostly are marine. *Hydra* is a freshwater hydrozoan inhabiting streams and lakes (Fig. 8.1a, b). It has a body size of

---

J.S. Hwang, S. Hayakawa, Y. Takaku, and T. Gojobori
Center for Information Biology and DDBJ, National Institute of Genetics, Mishima, 411-8540 Shizuoka, Japan
jhwang@lab.nig.ac.jp, shayakaw@lab.nig.ac.jp, ytakaku@lab.nig.ac.jp, tgojobor@genes.nig.ac.jp

S. Nagai
National Research Institute of Fisheries and Environment of Inland Sea, Maruishi 2-17-5, Hatsukaichi, 739-0452 Hiroshima, Japan
snagai@affrc.go.jp

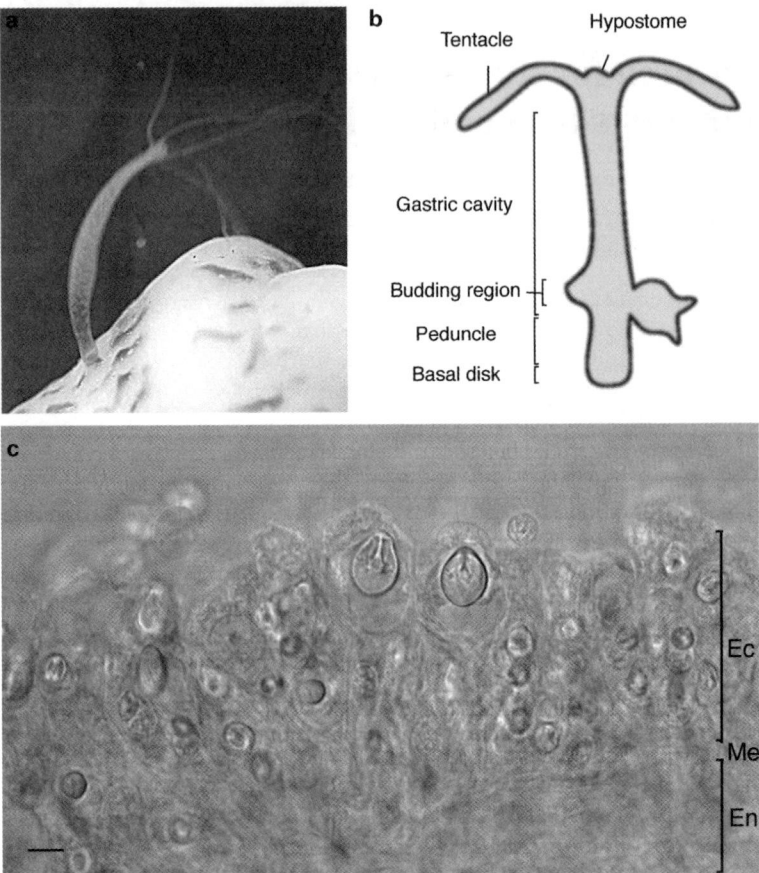

Fig. 8.1 a Freshwater polyp of *Hydra magnipapillata*. **b** *Hydra* and the terms used to describe the morphology of *Hydra*. **c** Ectodermal epithelial layer at the tentacle showing nematocytes mounted in the battery cells. *Ec* ectoderm, *Me* mesoglea, *En* endoderm. *Scale bar* 2 μm

5–15 mm and can undergo either sexual or asexual reproduction (i.e., budding). The life cycle of *Hydra* is simple compared with that of other cnidarians since the sexually reproduced *Hydra* remains as a polyp at all times without undergoing the planula and medusa stages. *Hydra* has been commonly used in the laboratory for more than 200 years and today it is still one of the famous experimental animals for research on regeneration, body plan organization, the stem-cell system, toxicity and evolution of metazoans.

8.1.1 A Stinging Cell Type in Cnidarians

All cnidarians possess a type of cells called nematocytes (or cnidocytes) (Hessinger and Lenhoff 1988; Kass-Simon and Scappaticci 2002). A nematocyst is a specialized organelle of a mature nematocyte (Figs. 8.1c, 8.2). So far more than

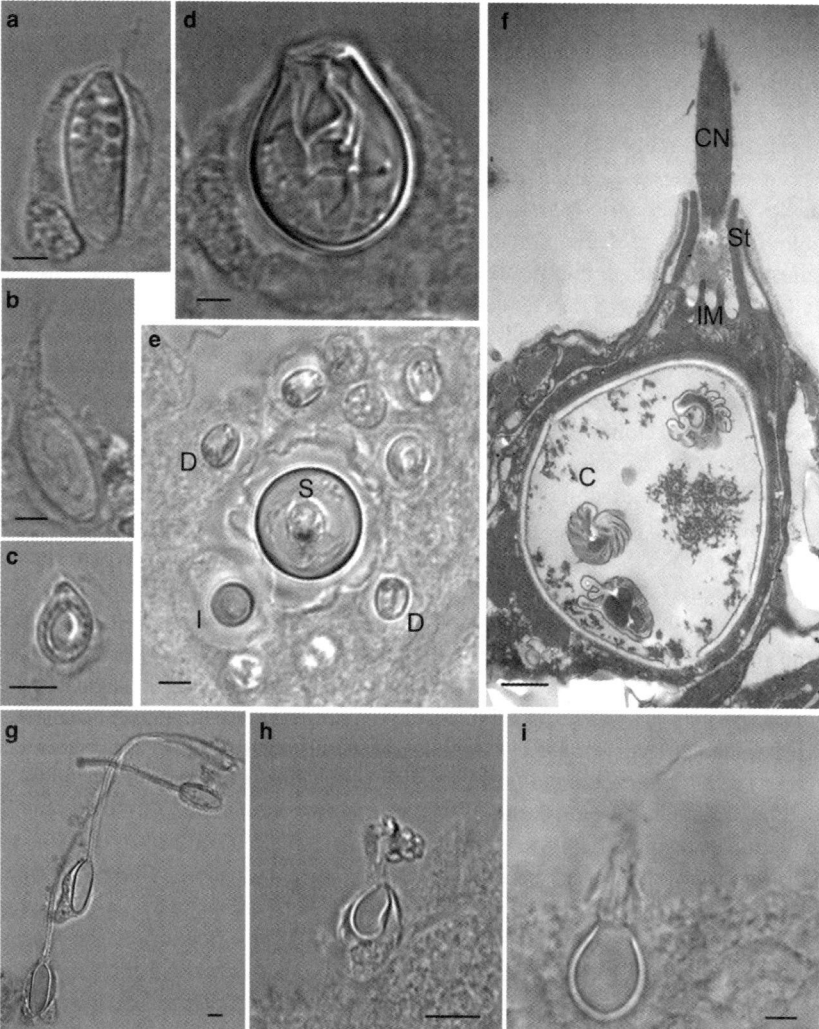

Fig. 8.2 Nematocytes of *Hydra*. **a–d** Four types of nematocytes include holotrichous isorhiza (**a**), atrichous isorhiza (**b**), desmoneme (**c**) and stenotele (**d**). **e** Top view of a tentacle shows a stenotele is surrounded by one isorhiza and eight to ten desmonemes in each battery cell. **f** Transmission electron micrograph of a stenotele. It is composed of a cnidocil apparatus at the apical surface of the tentacle and a capsule lying just beneath the cnidocil apparatus. **g–i** Discharged nematocysts: holotrichous isorhiza (**g**), desmoneme (**h**) and stenotele (**i**). *S* stenotele, *D* desmoneme, *I* isorhiza, *CN* cnidocil, *St* stereocilia or outer microvilli, *IM* inner microvilli, *C* capsule. *Scale bar* 1 μm (**a–f**), 2 μm (**g–i**)

25 structurally distinct nematocysts have been found in Cnidaria and their roles include prey capture, defense and adhesive locomotion. Most extensive studies of nematocysts have been accomplished in *Hydra*, which has four types of nematocysts, whereas parasitic *Polypodium* is known to have one type of nematocyst, the atrichous isorhiza (Raikova 1990). The four types of nematocytes in *Hydra*

are, namely, stenotele, desmoneme, holotrichous isorhiza and atrichous isorhiza (Chapman and Tilney 1959a, b; Fig. 8.2a–d). Nematocytes are interspersed between epithelial cells in the ectoderm (Fig. 8.1c). At the tentacles, eight to ten nematocytes are enveloped by a modified epithelial cell named a "battery cell," forming a nematocyte–battery cell complex (Campbell 1987; Fig. 8.2e). Among the four types of nematocytes, stenotele is the biggest in size and is distributed in both tentacles and the gastric region (Fig. 8.2d). Discharge of stenotele facilitates the capture of prey by propelling the toxin and paralyzing the prey. Desmoneme is the smallest nematocyst but contributes to 82% of total nematocysts in *Hydra*, mostly found in tentacles (Fig. 8.2c). It also functions in prey capture. Electron microscopy revealed that holotrichous isorhiza has a rod-shaped body and is predominantly used in defense by releasing poisonous substances (Fig. 8.2a). Atrichous isorhiza can be distinguished from holotrichous isorhiza by its smaller size and slightly oval shape (Fig. 8.2b). Unlike holotrichous isorhiza, all atrichous isorhizas migrate to the tentacles after the maturation (Bode and Flick 1976). Functionally, it releases a sticky substance for attachment to a surface during locomotion.

Nematocysts of *Hydra* consist of a hard-walled capsule and a ciliary cnidocil apparatus (Fig. 8.2f). The relaxed capsule is filled with a high concentration of poly(γ-glutamate) and toxins that create an intracapsular pressure. The high concentration of poly(γ-glutamate) provides a high osmotic pressure of 150 bar in the capsule and guarantees an explosive extrusion of the nematocyst (Weber 1990; Fig. 8.2g–i). Upon the chemical or mechanical stimulus, the cnidocil apparatus of the stenotele acts as a trigger which functions to generate two steps of the explosion process (Holstein and Tardent 1984). In the first step, the operculum is opened as a result of the high intracapsular pressure and the stylets are ejected. The ejection of stylets punches a hole in the tissue of the prey. This is immediately followed by the evagination of the long tubule and barbs. This second step of explosion involves the swelling of poly(γ-glutamate) inside the tubular lumen and the eversion of the tubular thread and hollow barbs. In this step the toxic substances can be delivered into the prey through the tubule and barbs (Lotan et al. 1995). The whole discharge process occurs in less than 3 ms (Holstein and Tardent 1984; Nüchter et al. 2006). Evidence also points out that the nematocyst discharge is calcium-dependent (Lubbock and Amos 1981; Gitter et al 1994; Watson and Hessinger 1994) and is also modulated by the nervous system (Brinkmann et al. 1996; Thurm et al. 2004).

8.1.2 Polykrikos and the Extrusive Organelles

Stinging weapons used for prey capture and defense are not only restricted to cnidarians, a variety of exclusive organelles are also commonly found in protists. These include trichocyst, mucocyst, toxicyst, haptocyst, rhabdocyst, nematocyst, etc. as described in Hausmann (1978). The trichocyst (also known as the secretory granule) of the ciliate *Paramecium* consists of a tip and a body and is attached to the cell membrane. It plays a role in defense and is discharged under mechanical or electrical stimulation (Harumoto and Miyake 1991). The discharged trichocyst is 22–24 µm in length, which is eightfold longer than the resting body (Bannister 1972). The elongation is due to the expansion of the filamentous matrix of the body part. Trichocysts are not only found in ciliated protozoans, flagellates also contain a trichocyst-like organelle (Wehrmeyer 1970; Bouck and Sweeney 1966). Another example of extrusome, toxicyst, has been reported in *Didinium*, *Loxophyllum* and *Litonotus* (Iwadate et al. 1999; Hausmann 1978; Rosati and Modeo 2003). In general, toxicysts of different ciliates have an ejected tubule and are capable of releasing toxic materials. *Dididium* feeding on *Paramecium* discharges toxicysts from its proboscis and *Paramecium* can be instantly immobilized and killed (Iwadate et al. 1999). *Litonotus* also releases toxicysts when it collides with the ciliates of prey. The prey soon stops moving as the locomotory cilia are deformed by the effect of toxicysts (Verni 1985). The predatory action continues with the specific recognition of prey by *Litonotus* and is then followed by engulfment (Rosati and Modeo 2003; Ricci and Verni 1988).

It has been recognized for some time by protozologists that some extrusive organelles of protists highly resemble the cnidarian nematocysts in morphology (Hausmann 1978). The ciliate *Remanella* contains a type of pear-shaped extrusive organelle. It structurally resembles the desmoneme of *Hydra* in which a long coiled filament is enclosed inside a capsule and an apical cap containing longitudinal fibrils (Raikov 1992). Even more surprisingly, the so-called nematocyst-taeniocyst of *Polykrikos* looks significantly like the structure of the stenotele and its developing form in *Hydra* (Greuet and Hovasse 1977; Westfall 1983; Fig. 8.3c, f). *Polykrikos* is a heterotrophic dinoflagellate found initially by a German zoologist, Otto Bütschli, in 1873. It is classified in the order Gymnodiniales, which is generally characterized by an unarmored form, a shallow apical groove, a longitudinal groove or sulcus, a transverse furrow or cingulum and two flagella, one emerges from the sulcus and the other from the cingulum (Fig. 8.3a, b). The free-living form of *Polykrikos* is a pseudocolony containing two to eight zooids and about half the number of nuclei over its zooids (Nagai 2002; Fig. 8.3d, e). The nematocyst of *Polykrikos* has a thick-walled capsule that contains a long, everted thread attached to a big stylet at one end and an opening cap. The current knowledge of the *Polykrikos* nematocyst is only based on studies conducted by a handful of researchers and, among them, Westfall et al. (1983) have reported the ultrastructure of nematocysts using transmission electron microscopy. In addition to nematocysts, *Polykrikos* contains other types of extrusome, trichocyst and mucocyst (Hoppenrath and Leander 2007).

8.2 Comparison of Nematocysts between *Hydra* and *Polykrikos*

8.2.1 Similarities in the Structure of Nematocysts between Hydra and Polykrikos

Westfall et al. (1983) have observed various developmental stages of the nematocyst-taeniocyst in *Polykrikos kofoidii*. The nematocyst-taeniocyst is produced by the secretion of Golgi complexes. The secretory products are continually deposited

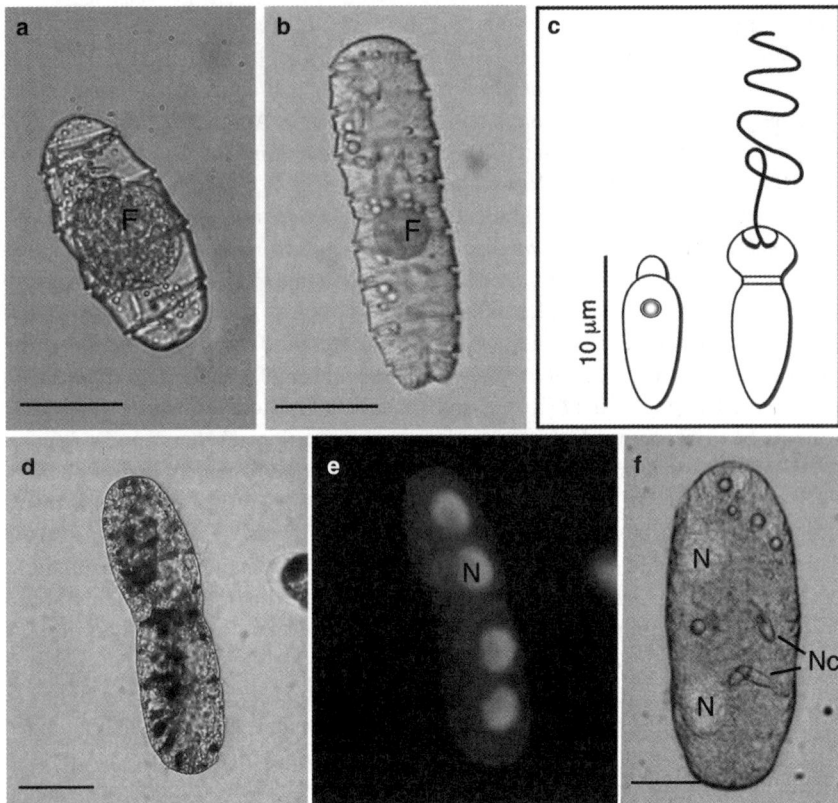

Fig. 8.3 Pseudocolony of *Polykrikos kofoidii*. **a** A pseudocolony of *Polykrikos kofodii* is characterized by two, four or eight zooids, approximately 80 – –100 μm in length and 30-50 μm in diameter. **b** A pseudocolony undergoes longitudinal fission. **c** Relaxed and discharged nematocysts of *Polykrikos kofodii*. **d, e** When stained with 4′,6-diamidino-2-phenylindole, four nuclei are observed in a pseudocolony. **f** A fixed pseudocolony contains numerous nematocysts. *F* food, *N* nucleus, *Nc* nematocyst. *Scale bar* 30 μm

inside the capsule of the nematocyst, forming a thick capsule wall, a stylet and a long, hollow tube. Compared with the differentiation of the *Hydra* (or cnidarian) nematocyst, the specialized structures of two different organisms appeared to originate from the rough endoplasmic reticulum and the Golgi apparatus. Yet, in *Hydra*, the filamentous tubule is first synthesized within the cytoplasm of the nematocyte and later invaginates into the capsule. Although the developmental steps of the nematocyst are not completely the same, both share structural similarities in the mature nematocyst. In *Polykrikos*, the mature nematocyst disconnects from the taeniocyst but both are still held in an enclosed chute (Figs. 14, 24 in Westfall et al. 1983). The nematocyst-taeniocyst chute is positioned near the sulcus and flagella. The most striking similarities between the *Polykrikos* nematocyst and the *Hydra* stenotele are the filamentous tubule and the stylet formed within the capsule (Fig. 8.4a, b). In

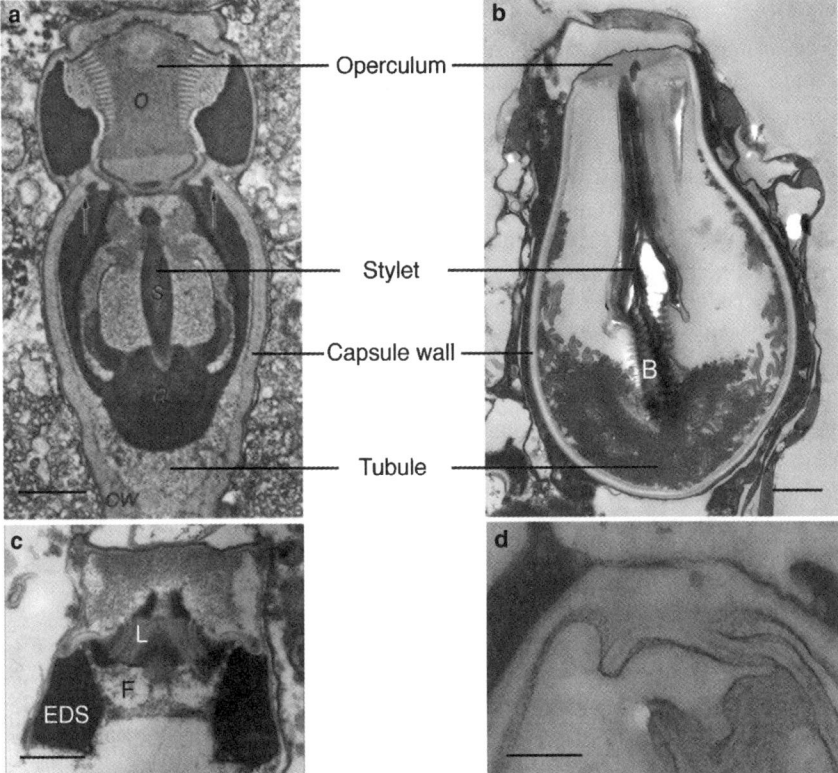

Fig. 8.4 Structural comparison between *Polykrikos* and *Hydra* nematocysts. **a** *Polykrikos* nematocyst and this figure is adapted from Westfall et al (1983). **b** *Hydra* nematocyst. **c** *Polykrikos* operculum. **d** *Hydra* operculum. *o* operculum, *s* stylet, *a* anterior chamber, *cw* capsule wall, *B* barb, *L* lamellae, *EDS* electron-dense sheath, *F* fibrous material. *Scale bar* 1 μm

the *Polykrikos* nematocyst, the stylet, with one end pointed toward the operculum, has the other end attached to the anterior chamber and is aligned along the longitudinal axis of the capsule. The stylet of the *Hydra* stenotele also lies parallel to the longitudinal axis of the capsule with the tip pointed toward the operculum. The filamentous tubule of the *Polykrikos* nematocyst is positioned distal to the operculum and is connected to the anterior chamber inside the capsule; whereas the tubule of the stenotele is an extension of the stylet and is tightly coiled in the capsule (Fig. 8.4a, b). Nevertheless, the nematocyst tubules of both organisms are arranged in an inverted form inside the capsule. The operculum, a lid that closes the capsule, is identified in the nematocysts of both *Polykrikos* and *Hydra*. However, they may differ in detail. In the *Hydra* stenotele, the operculum has a uniformly laminar structure and the Material that constitutes this structure is also found in spines and the stylet (Koch et al. 1998; Fig. 8.4d). On the other hand, the *Polykrikos* operculum is more elaborate in structure. Transmission electron microscopy shows that the operculum has a thick lamellar structure surrounded by an electron-dense sheath. A fibrous material fills up the cavity inside the electron-dense sheath (Fig. 8.4c).

Since *Polykrikos* and *Hydra* share similarities in terms of nematocyst structure, the nematocysts are compared regarding their roles in feeding behavior. The use of two nematocysts of *Hydra*, stenotele and desmoneme, for catching prey has been well documented (Ewer 1947; Ruch and Cook 1984). Briefly, when prey is approaching *Hydra*, the stenotele can be mechanically discharged within milliseconds and the explosive tubule filament, stylet and numerous spines then penetrate into the cellular tissue of the prey (Thorington and Hessinger 1988; Watson and Hessinger 1989; Fig. 8.5a); simultaneously toxin within the capsule is released into the prey. The desmoneme fires its spiral tubule during discharge, is accompanied by the stenotele and acts as a trap to ensnare prey. Once the prey has been immobilized, it is inserted into the gastric cavity by tentacles through the mouth. Interestingly, when we observe the feeding behavior of *Polykrikos* with its prey, *Cochlodinium* sp., in a cultivated medium, *Polykrikos* is able to demonstrate a feeding behavior similar to that of *Hydra*. This feeding behavior has also been previously reported (Matsuoka et al. 2000).

8.2.2 The Strategy to Capture Prey

The feeding behavior of *Polykrikos* was observed under a microscope with ×40 phase-contrast water-immersion optics and recorded with a video camera (Fig. 8.5b–h). In general, the nematocyst is discharged from the living zooid, anchors and pulls its prey toward the posterior sulcus (Fig. 8.5d, e). Once *Polykrikos* gets in contact with the prey, it engulfs and ingests the entire prey along the sulcus (Fig. 8.5f–h). Our observation reveals four features of the *Polykrikos* nematocyst that extremely resemble those of the *Hydra* nematocyst in terms of prey capture:

1. Release of the nematocyst is triggered by mechanical stimulus. *Polykrikos* swims in a spiraling manner and ejects its nematocyst when it has physical contact with prey (Fig. 8.5d). The taeniocyst is thought to serve as a mechanical sensor for the discharge of the nematocyst in *Polykrikos* (Westfall et al. 1983).
2. The process of nematocyst discharge occurs in less than 1 second. In the case of *Hydra*, the discharge of the nematocyst requires only in a few milliseconds. Currently, we are not able to record the time on a scale of milliseconds, however, since the ejection of the trichocyst from *Paramecium* takes placed within milliseconds (Knoll et al. 1991) and we assume that the discharge of the *Polykrikos* nematocyst also occurs in a similar manner.
3. The stylet and thread are everted during nematocyst discharge. During the formation of nematocysts in *Polykrikos* and *Hydra*, the stylet and filamentous tubule are assembled in an inverted form in the capsule. The discharge process allows the content of the capsule to be everted and pierce the prey.

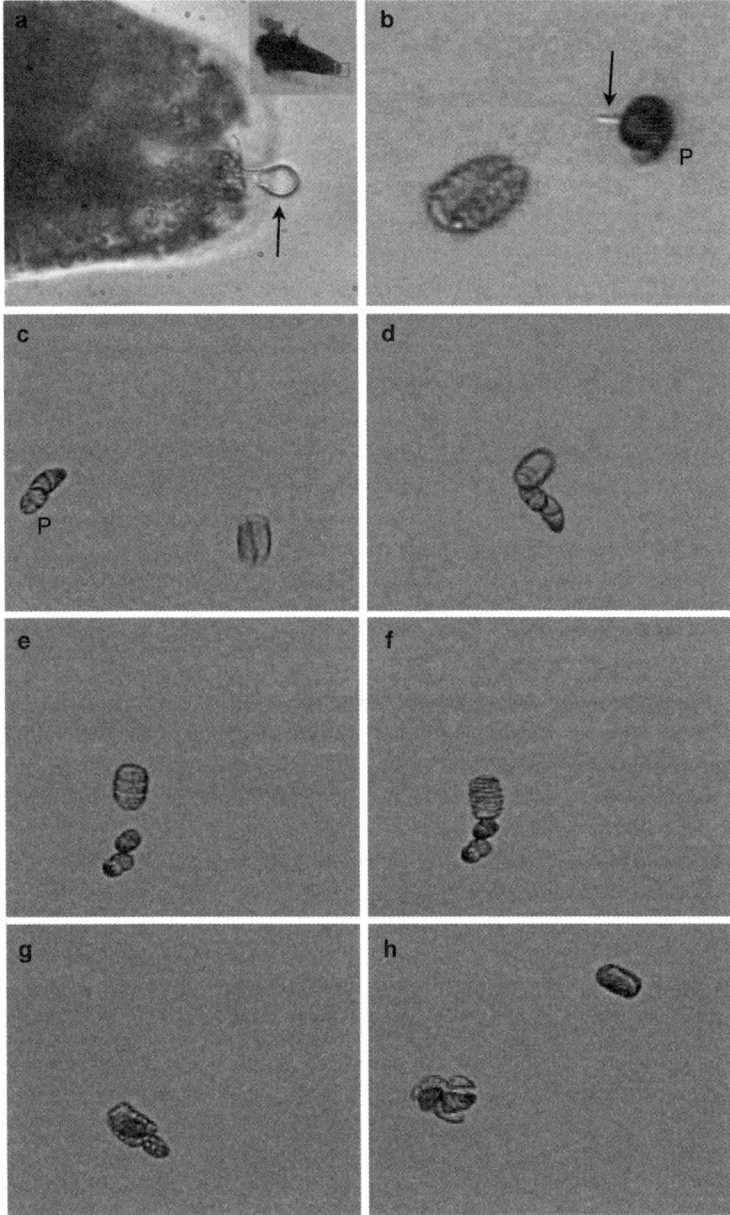

Fig. 8.5 Prey capture by *Polykrikos*. **a** A *Hydra* stenotele (*arrow*) is injected into a brine shrimp (*inset*). **b** A discharged nematocyst (*arrow*) of *Polykrikos* is anchored to the prey (*P*). **c** A free-swimming *Polykrikos* is approaching the prey (*P*). **d** The nematocyst discharge is triggered immediately after *Polykrikos* comes into contact with the prey. The stylet enables the nematocyst to penetrate the prey. **e** *Polykrikos* anchors the prey with a long tubule connecting to the nematocyst. **f** The tubule is shortened and *Polykrikos* pulls the prey back to its posterior sulcus. **g** *Polykrikos* engulfs the prey. **h** The prey is completely swallowed by *Polykrikos* and as a result the shape of *Polykrikos* is altered

4. The prey is paralyzed soon after the nematocyst penetrates the tissue. Lethal toxins have been identified stored inside the capsule of the *Hydra* stenotele and are released at the deep puncture wound (Tardent and Holstein 1982). Although no toxin has been reported in the nematocyst of *Polykrikos*, the observation on prey capture often shows that the prey stops moving soon after the nematocyst has been discharged. Therefore, the nematocyst may be able to release toxic substances and immobilize the prey.

Nevertheless, the *Polykrikos* nematocyst differs from the *Hydra* stenotele in one aspect regarding prey capture. The evaginated tubule of the *Hydra* stenotele during the discharge penetrates into the prey, making a hole that allows toxin to be delivered into the prey (Tardent and Holstein 1982). However, the tubule connecting *Polykrikos* and the injected nematocyst is used to pull the prey toward the predator by shortening its length (Fig. 8.5e). The mechanism of this shortening process is still unclear. It is speculated that there is another tubule connected to one end of the nematocyst proximal to *Polykrikos* and that is capable of anchoring the prey (Matsuoka et al. 2000).

8.3 Nematocyst-Related Genes

Both *Hydra* and *Polykrikos* have elaborated nematocysts. Formation of such a complex organelle requires at least hundreds of genes. Studies of *Hydra* nematocysts have identified several genes related to the skeletal morphology; these are the genes responsible for making the capsule wall, stylet and operculum. Yet, there are more genes involved in making these different components of the nematocyst since over 50 nematocyst-specific genes have been identified by complementary DNA microarray and at least two thirds of them are important for the assembly of the capsule wall, tubule filament, stylet, spines, operculum, cnidocil apparatus and numerous toxins (Hwang et al. 2007). Proteins coded by these genes have no sequence homology in the current protein database and they are classified as "cnidarian-specific." However, there are still many protists whose genomes have not yet been sequenced, and we may be able to find counterparts for the cnidarian-specific genes once the protist genomes, especially ciliate genomes, become more and more available. Of the remaining one third of the nematocyst proteins whose homologues are found in the database, a few are closely related to those of proteobacteria, plants or protozoans rather than the bilaterians. For example, two newly identified enzymes in the *Hydra* nematocyst (hydmg_006secrev_019 and hmp_06864 in Table 1 in Hwang et al. 2007) are highly similar to proteobacteria enzymes. It is also surprising that a calcium-binding protein of the nematocyst (hmp_10958) has no sequence homology in other organisms except *Trypanosoma* or *Plasmodium*. This indicates that at least part of the molecular basis of cnidarian nematocysts can be derived from an ancient origin.

8.3.1 Common Cytoskeletal Genes of Nematocysts

To mediate the release of the nematocyst, the nematocyst must be organized by a complex cytoskeletal network. In *Hydra*, common cytoskeletal proteins such as actin, the microtubule and the intermediate filament have been shown to take part in the assembly of the nematocyst and also to play a role in maintaining an apical, upright position of the nematocyte–battery cell complex in the tentacle (Wood and Novak 1982; Schertenleib and Stidwill 1991; Golz 1995). With use of antibodies against actin, tubulin and centrin, fluorescent signals could be observed at the apical region as well as the cytoplasm of the nematocyte (Golz 1995). Microfilaments (i.e., actin) are detected in stereocilia (or outer microvilli) of the cnidocil apparatus. These stereocilia of the nematocyst are comparable to the stereocilia found in the hair cells of the vertebrate inner ear or the lateral line of fish in terms of receiving environmental stimuli via chemo- or mechanotransduction (Tilney and Tilney 1992). Microtubules are also parts of the cnidocil apparatus in which the axoneme of the cnidocil is densely stained with the tubulin antibody (Golz 1995). Moreover, the capsule is tightly surrounded by a microtubule framework which is commonly known as a microtubular basket. Intermediate filaments are localized in the cytoplasm with close proximity to the capsule and form a filamentous meshwork associated with the basal nematocyte junction adjacent to the battery cell (Wood and Novak 1982). Both the microtubular basket and the intermediate filaments serve to anchor and support the nematocyst in a "fire-ready" position.

In order to determine the molecular components of the cytoskeletal elements in the *Polykrikos* nematocyst, we tested *Polykrikos* with anti-β-tubulin antibody and rhodamine phalloidin. The anti-β-tubulin staining shows the longitudinal microtubules running in anterior–posterior direction in the theca (Fig. 8.6a). These longitudinal microtubules are aligned parallel to one another with space between adjacent microtubules. There are also longitudinal microtubules densely packed in the circulum (or girdle), showing a transverse band under the confocal microscope (Fig. 8.6a). The microtubule also constitutes the "$9+2$" pattern of the axoneme in the flagellum (Gibbons 1981; Mitchell 2004), a major characteristic of a dinoflagellate. A microtubule was not found in the nematocyst of *Polykrikos* when an antibody that was originally raised against β-tubulin of sea urchin was used. On the other hand, actin was detected in the operculum of the nematocyst by using rhodamine phalloidin (Fig. 8.6b). The operculum of the nematocyst has a complex structure as described above and its structure seems to be more elaborate than that of *Hydra*. The presence of actin should be critical for the opening of the operculum from a closed conformation. During the discharge of the nematocyst, the triggered operculum opens to release heterogeneous content of the capsule. It is not known how the complex operculum is opened since all actions are completed in milliseconds. In the case of *Hydra*, the sequences that lead to the opening of operculum were not observed even though the discharge process is recorded in nanoseconds (Nüchter et al. 2006). No actin has been previously reported in the operculum of nematocysts in cnidarians.

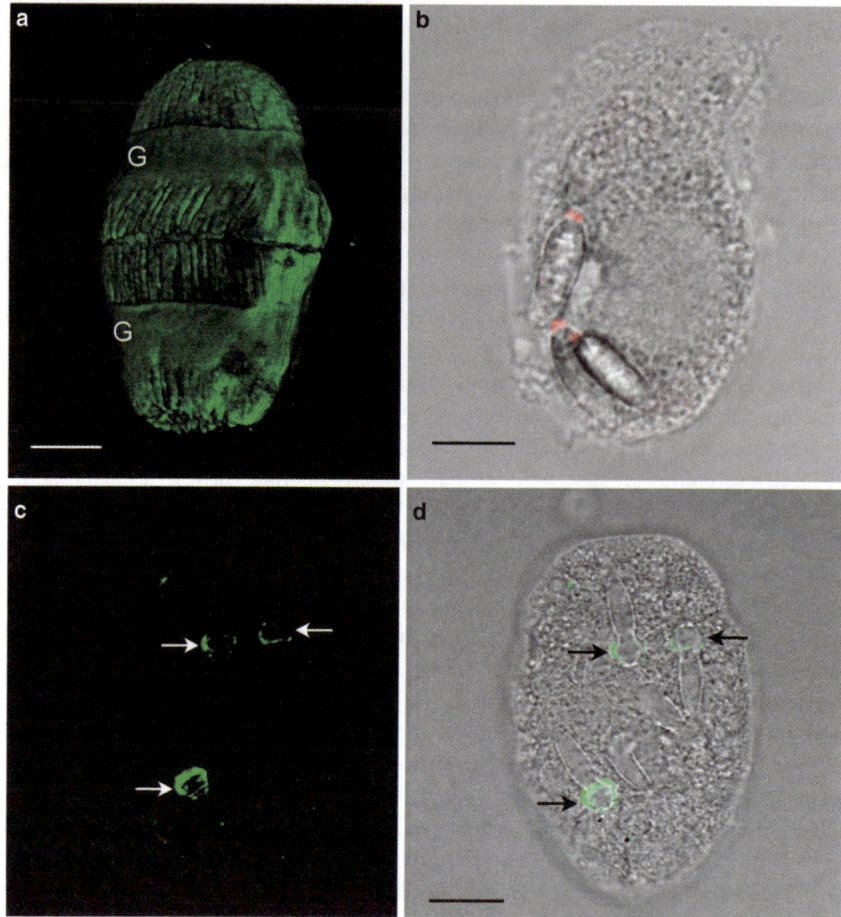

Fig. 8.6 Localization of cytoskeletal proteins in *Polykrikos* by immunofluorescence. **a** A 3D image of a single zooid is labeled with anti-β-tubulin antibody. β-Tubulin (*green*) running longitudinally along the cell axis is observed beneath the theca. The fluorescent signal of β-tubulin is stronger in the girdle (*G*) than in any other regions. **b** Positive phalloidin labeling (*red*) is clearly visible in the operculum of the nematocyst. **c, d** A polyclonal antibody against minicollagen in the capsule wall of a *Hydra* nematocyst is used to label *Polykrikos*. The signal (*green*) is mainly detected at the anterior chamber of the nematocyst (*arrow*). *Scale bar* 10 μm

8.3.2 Nematocyst-Specific Genes in Hydra are not found in other Metazoans

Apart from the conserved cytoskeletal genes, *Hydra* express another set of cytoskeletal genes which share no homology with those of other metazoans. The first nematocyst gene is identified to code for the capsule wall protein named "minicollagen" (Kurz et al. 1991; Holstein et al. 1994). Minicollagen contains the

Gly-X-Y repeat as seen in the collagen family but it is much shorter than collagen. N- and C-termini have a cysteine-rich domain in which cysteine can make intramolecular disulfide bonds. These disulfide bonds are important for the polymerization and condensation of minicollagen at the inner capsule wall (Engel et al. 2001). Although minicollagen shares similar domain structures with collagen, it is a cnidarian-specific protein and not an orthologue of collagen since collagen types I and IV have been reported in *Hydra* (Deutzmann et al. 2000; Fowler et al. 2000). So far more than 15 minicollagen genes have been identified from *Hydra* expressed sequence tags and genomic scaffolds (Hwang et al. 2007). NOWA is another cnidarian-specific gene responsible for forming the outer capsule wall of the nematocyst (Engel et al. 2002). It has a C-type lectin-like domain and a rodent sperm-coating glycoprotein domain at the N-terminal half of the protein. Interestingly, its C-terminus is a cysteine-rich domain which closely resembles that of minicollagen at C- and N-termini, and it is suggested that both NOWA and minicollagen are linked by disulfide bonds during the assembly of the capsule wall (Özbek et al. 2004). A gene that codes for the spine and operculum protein, spinalin, has also been identified in *Hydra* (Koch 1998). It has several putative domains with no homology in the database. All three nematocyst proteins have a short sequence of signal peptide at the N-terminus, indicating they are transported through the endoplasmic reticulum and processed in the Golgi apparatus.

8.3.3 Do Hydra and Polykrikos Share Homologous Protein in Making Nematocysts?

We have compared the nematocysts of *Polykrikos* and *Hydra* based on the morphological similarities and showed that each detailed structure of the former has its corresponding part in the later. In this section, we try to determine the existence of minicollagen in the nematocyst of *Polykrikos*. Minicollagen 1 constitutes the inner wall of the capsule in *Hydra* and is a taxon-specific protein without an orthologue in other metazoans. We applied the antibody of minicollagen 1 (gift of T. Holstein and S. Özbek, University of Heidelberg) to *Polykrikos* fixed by Bouin's fixative. The antibody stains the surface of the anterior portion of the discharged nematocyst as shown in Fig. 8.6c and d. Bouin's fixative causes many *Polykrikos* nematocysts to be discharged inside the cell body (Harumoto and Miyake 1991) and those having evaginated anterior chamber (Fig. 8.6c, d, arrows) are labeled with fluorescent dye. The nondischarged nematocysts show no significant labeling of fluorescence. The failure of the antibody to stain nondischarged nematocysts may be due to the poor penetration of the antibody into the capsule. A nondischarged capsule is closely sealed with an operculum and its doubled wall has low permeability for the antibody. No fluorescent signal is detected in any other regions in *Polykrikos* except the discharged nematocysts. The result suggests that the anterior chamber of the nematocyst is composed of minicollagen-like material or the protein having at least an epitope that is identical to an epitode of the minicollagen.

8.4 Evolution of Nematocysts?

Nematocysts are not commonly found in metazoans, and their existence is only restricted to the phylum Cnidaria. The most basal metazoan, sponge, produces neither nematocysts nor similar extrusive structures. It has been reported that the nemertean worms, especially the class Enopla, have a specialized stylet apparatus for attacking prey (Stricker and Cloney 1981). The stylet apparatus is stored in the proboscis chamber. When the proboscis is everted, the central stylet punches a hole into the prey tissue and neurotoxin is then released into the lesion to immobilize the prey. Although the nemertean uses a similar prey-capturing strategy as seen in cnidarians, the precise structure of the stylet apparatus is quite different from the nematocysts of cnidarians. The major structural differences observed in the nemertean stylet apparatus include no rigid capsule that envelops the stylet, the lack of a filamentous thread and the presence of calcified material at the outer cortex of the stylet shaft (Stricker and Cloney 1982; Stricker 1985). Nevertheless, the discovery of the *Polykrikos* nematocyst may be evidence against the likelihood that the nematocyst is a taxon-specific organelle. In our study, we demonstrated similarities of the nematocysts of *Hydra* and *Polykrikos* with regard to the feeding behavior, the morphology and the molecular basis. We discuss in the next section whether these similarities are the results of convergence in evolution or the result of a common origin.

8.4.1 Gene Loss or Lateral Gene Transfer?

So far, the number of genes involved in the differentiation pathway of cnidarian nematocytes is greater than 50, and there should be at least 30 or more specialized for coding cytoskeletal proteins (Hwang et al. 2007). Almost all cytoskeletal genes of nematocysts identified so far have no orthologue in other organisms. One may think that the nematocyst occurrence is due to convergent evolution, that is, all nematocyst-responsible genes emerged after the divergence between cnidarians and bilaterians, and continued to evolve in the lineage of cnidarians until the present day. Yet, how did nematocyst cytoskeletal genes emerge in the last common ancestor of cnidarians? One possible clue is provided by the study of a sea anemone genome in which the "new" genes arose in the eumetazoan ancestor via the mechanism of domain shuffling or domain recombination of preexisting genes (Putnam et al. 2007). In addition, gene duplication might be involved but it has to be followed by a large sequence divergence. To support the convergence hypothesis, further investigation is necessary to reveal the molecular materials that constitute the structure and the differentiation pathway of nematocysts in both cnidarians and *Polykrikos*.

On the other hand, our studies strongly support a common origin of nematocysts for cnidarians and *Polykrikos*. The absence of nematocysts in other animals could be possibly due to three evolutionary causes: (1) most nematocyte-specific genes were lost in the bilaterian lineage after the divergence between Cnidaria and other animals; (2) genes were laterally transferred from protoctista to the lineage of Cnidaria;

and (3) genes were transferred to Cnidaria via the symbiotic relationship. The first hypothesis requires a massive gene loss in many taxa, considering those genes are quite important for the assembly of a nematocyst. Moreover, the lack of nematocysts in sponge, which is a more primitive multicellular organism than *Hydra*, implies that a double gene loss happened in both sponge and bilaterian lineages. Searching for the genome of a demosponge, *Amphimedon queenslandica*, failed to detect the homologous sequences of most nematocyst genes of *Hydra* (Hwang et al. 2007). It is not known for sure whether lateral gene transfer or symbiosis supports the hypothesis of a common origin of nematocysts. In fact, the endosymbiotic relationship is common between Cnidaria and Protoctista. The dinoflagellate genus *Symbiodinium* is a well-known symbiont of most corals, sea anemones and jellyfish (Baker 2003; Fitt and Trench 1983; Lajeunesse and Trench 2000). Moreover, green *Hydra* such as *H. viridissima* and *H. circumcincta* harbor *Chlorella* in the endodermal layer as a symbiont and these green *Hydra* are thought to be more ancestral forms than other *Hydra* (Zacharias et al. 2004). Therefore, it is hypothesized that the ancestral form of the nematocyst belongs to a symbiont of cnidarians and subsequently nematocyst genes were extensively transferred from the symbiont to the host nucleus. Particularly, symbiosis is one of the evolutionary resources that contributes to the novelty of genes and the diversity of species. Shostak (1993) has proposed a hypothesis to explain the coincidence of having the nematocyst-like organelle in both cnidarians and some protozoans. According to his hypothesis, cnidarians are the consequence of two (or more) eukaryotic cells coming to stay as mutual symbionts and eventually their nuclei could have fused and undergone nuclear merger. Furthermore in his hypothesis, epithelial cells of *Hydra* could have derived from one kind of eukaryotic cell and the interstitial cell, the precursor of the nematocyte and the neuron, comes from the protozoan. However, a paper by Müller et al. (2004) provides evidence that the interstitial cells in *Hydractinia* possess totipotency and are capable of differentiating into not only nematocytes, nerve cells, gland cells and gametes (a similar process as seen in *Hydra*), but also epithelial cells. The transition of interstitial cells into epithelial cells in *Hydractinia* suggests one origin of hydroid cells, rather than two different origins of cell types that were described by Shostak (1993). Perhaps the ancestral nematocyst of cnidarians arose from an endosymbiosis relationship with a dinoflagellate containing a nematocyst-like organelle and subsequently the nematocyst-like organelle relinquished many genes to the host genome, thereby enriching both the size of the nuclear genome and the plasticity of the cell types of cnidarians. A series of endosymbiosis followed by gene transfer would have had a great effect on eukaryote evolution and would have introduced novelty and complexity to current eukaryotic cells (Smith 1989; Dyall et al. 2004). The invertebrate–algae symbiosis is also a widespread event but its evolutionary significance has been documented less so far. The most commonly found invertebrate for a algal symbiont is in cnidarians and their symbiotic relationship is often dominated in both marine and freshwater environments (Rahat 1985; Goodson et al. 2001; Belda-Baillie et al. 2002). It is an interesting issue whether the cnidarian–algae symbioses would have played a key role in the evolutionary changes that brought new features and novelties to the phylum Cnidaria. Perhaps this question can be resolved in future studies.

References

Baker AC (2003) Flexibility and specificity in coral-algal symbiosis: diversity, ecology, and biogeography of *Symbiodinium*. Annu Rev Ecol Evol Syst 34:661–689

Bannister LH (1972) The structure of trichocysts in *Paramecium caudatum*. J Cell Sci 11:899–929

Belda-Baillie CA, Baillie BK, Maruyama T (2002) Specificity of a model cnidarian-dinoflagellate symbiosis. Biol Bull 202:74–85

Bode HR, Flick KM (1976) Distribution and dynamics of the nematocytes populations in *Hydra attenuata*. J Cell Sci 21:15–34

Bouck GB, Sweeney BM (1966) The fine structure and ontogeny of trichocysts in marine dinoflagellates. Protoplasma 61:205–223

Brinkmann M, Oliver D, Thurm U (1996) Mechanoelectric transduction in nematocytes of a hydropolyp (Corynidae). J Comp Phys 178:125–138

Campbell RD (1987) Organization of the nematocyst battery in the tentacle of hydra: arrangement of the complex anchoring junctions between nematocytes, epithelial cells, and basement membrane. Cell Tissue Res 249:647–655

Chapman GB, Tilney LG (1959a) Cytological studies of the nematocysts of Hydra: I. Desmonemes, isorhizas, cnidocils, and supporting structures. J Cell Biol 5:69–77

Chapman GB, Tilney LG (1959b) Cytological studies of the nematocysts of Hydra: II. The stenoteles. J Cell Biol 5:79–83

Deutzmann R, Fowler S, Zhang X, Boone K, Dexter S, Boot-Handford RP, Rachel R, Sarras MP (2000) Molecular, biochemical and functional analysis of a novel and developmentally important fibrillar collagen (Hcol-I) in hydra. Development 127:4669–4680

Dyall SD, Brown MT, Johnson PJ (2004) Ancient invasions: from endosymbionts to organelles. Science 304:253–257

Engel U, Pertz O, Fauser C, Engel J, David CN, Holstein TW (2001) A switch in disulfide linkage during minicollagen assembly in *Hydra* nematocysts. EMBO J 20:3063–3073

Engel U, Özbek S, Streitwolf-Engel R, Petri B, Lottspeich F, Holstein TW (2002) Nowa, a novel protein with minicollagen Cys-rich domains, is involved in nematocyst formation in *Hydra*. J Cell Sci 115:3923–3934

Ewer RF (1947) On the functions and mode of action of the nematocysts of hydra. Proc Zool Soc Lond 117:365–376

Fitt WK, Trench RK (1983) Endocytosis of the symbiotic dinoflagellate *Symbiodinium microadriaticum* Freudenthal by endodermal cells of the scyphistomae of *Cassiopeia xamachana* and resistance of the algae to host digestion. J Cell Sci 64:195–212

Fowler SJ, Jose S, Zhang X, Deutzmann R, Sarras Jr. MP, Boot-Handford RP (2000) Characterization of hydra type IV collagen. J Biol Chem 275:39589–39599

Gibbons IR (1981) Cilia and flagella of eukaryotes. J Cell Biol 91:107–124

Gitter AH, Oliver D, Thurm U (1994) Calcium- and voltage-dependence of nematocyst discharge in *Hydra vulgaris*. J Comp Phys 175:115–122

Golz R (1995) Cytoskeletal elements in migrating and tentacle-integrated nematocytes of a marine hydrozoan. Biol Cell 83:49–59

Goodson MS, Whitehead LF, Douglas AE (2001) Symbiotic dinoflagellates in marine Cnidaria: diversity and function. Hydrobiologia 461:79–82

Greuet C, Hovasse R (1977) A propos de la genese des nematocysts de *Polykrikos schwartzi* Bütschli. Protistologica 1:145–149

Harumoto T, Miyake A (1991) Defensive function of trichocysts in *Paramecium*. J Exp Zool 260:84–92

Hausmann K (1978) Extrusive organelles in protists. Int Rev Cytol 52:197–276

Hessinger DA, Lenhoff HM (1988) The biology of nematocysts. Academic, San Diego

Holstein T, Tardent P (1984) An ultrahigh-speed analysis of exocytosis: nematocyst discharge. Science 223:830–833

Holstein TW, Benoit M, v Herder G, Wanner G, David CN, Wanner G, Gaub HE (1994) Fibrous mini-collagens in *Hydra* nematocysts. Science 265:402–404

Hoppenrath M, Leander BS (2007) Morphology and phylogeny of the pseudocolonial dinoflagellates *Polykrikos lebourae* and *Polykrikos herdmanae* n. sp. Protist 158:209–227

Hwang JS, Ohyanagi H, Hayakawa S, Osato N, Nishimiya-Fujisawa C, Ikeo K David CN, Fujisawa T, Gojobori T (2007) The evolutionary emergence of cell type-specific genes inferred from the gene expression analysis of *Hydra*. Proc Natl Acad Sci USA 104:14735–14740

Iwadate Y, Katoh K, Kiuyama M, Asai H (1999) Ca^{2+} triggers toxicyst discharge in *Didinium nasutum*. Protoplasma 206:20–26

Kass-Simon G, Scappaticci AA (2002) The behavioral and developmental physiology of nematocysts. Can J Zool 80:1772–1794

Knoll G, Braun C, Plattner H (1991) Quenched flow analysis of exocytosis in *Paramecium* cells: time Course, changes in membrane structure, and calcium requirements revealed after rapid mixing and rapid freezing of intact cells. J Cell Biol 113:1295–1304

Koch AW, Holstein TW, Mala C, Kurz E, Engel J, David CN (1998) Spinalin, a new glycine- and histidine-rich protein in spines of *Hydra* nematocysts. J Cell Sci 111:1545–1554

Kurz EM, Holstein TW, Petri BM, Engel J, David CN (1991) Mini-colagens in hydra nematocytes. J Cell Biol 115:1159–1169

Lajeunesse TC, Trench RK (2000) Biogeography of two species of *Symbiodinium* (Freudenthal) inhabiting the intertidal sea anemone *Anthopleura elegantissima* (Brandt). Biol Bull 199:126–134

Lotan A, Fishman L, Loya Y, Zlotkin E (1995) Delivery of a nematocyst toxin. Nature 375:456

Lubbock R, Amos WB (1981) Removal of bound calcium from nematocyst contents causes discharge. Nature 290:500–501

Matsuoka K, Cho HJ, Jacobson DM (2000) Observation of the feeding behavior and growth rates of the heterotrophic dinoflagellate *Polykrikos kofoidii* (Polykrikaceae, Dinophyceae). Phycologia 39:82–86

Mitchell DR (2004) Speculations on the evolution of 9+2 organelles and the role of central pair microtubules. Biol Cell 96:691–696

Müller WA, Teo R, Frank U (2004) Totipotent migratory stem cells in a hydroid. Dev Biol 275:215–224

Nagai S, Matsuyama Y, Takayama H, Kotani Y (2002) Morphology of *Polykrikos kofoidii* and *P. schwartzii* (Dinophyceae, Polykrikaceae) cysts obtained in culture. Phycologia 41:319–327

Nüchter T, Benoit M, Engel U, Özbek S, Holstein T (2006) Nanosecond-scale kinetics of nematocyst discharge. Curr Biol 16:R316–R318

Özbek S, Pokidysheva E, Schwager M, Schulthess T, Tariq N, Barth D, Milbradt AG, Moroder L, Engel J, Holstein TW (2004) The glycoprotein NOWA and minicollagens are part of a disulfidelinked polymer that forms the cnidarian nematocyst wall. J Biol Chem 279:52016–52023

Putnam NH, Srivastava M, Hellsten U, Dirks B, Chapman J, Salamov A, Terry A, Shapiro H, Lindquist E, Kapitonov V, Jurka J, Genikhovich G, Grigoriev IV, Lucas SM, Steele RE, Finnerty JR, Technau U, Martindale MQ, Rokhsar DS (2007) Sea anemone genome reveals ancestral eumetazoan gene repertoire and genomic organization. Science 317:86–94

Rahat M (1985) Origin and evolution of symbiosis in green hydra. Arch Sci 38:385–399

Raikov IB (1992) Unusual extrusive organelles in karyorelictid ciliates: an argument for the ancient origin of this group. BioSystems 28:195–201

Raikova EV (1990) Fine structure of the nematocytes of *Polypodium hydriforme* Ussov (Cnidaria). Zool Scr 19:1–11

Ricci N, Verni F (1988) Motor and predatory behavior of *Litonotus lamella* (Protozoa, Ciliata). Can J Zool 66:1973–1981

Rosati G, Modeo L (2003) Extrusomes in ciliates: diversification, distribution, and phylogenetic implications. J Eukaryot Microbiol 50:383–402

Ruch RJ, Cook CB (1984) Nematocyst inactivation during feeding in *Hydra littoralis*. J Exp Biol 111:31–42

Schertenleib U, Stidwill RP (1991) The differentiation of cytoskeletal structures in nematocytes of *Hydra*. Hydrobiologia 216/217:678–684

Shostak S (1993) A symbiogenetic theory for the origins of cnidocysts in Cnidaria. BioSystems 29:49–58
Smith JM (1989) Generating novelty by symbiosis. Nature 341:284–285
Stricker SA (1985) The stylet apparatus of *Monostiliferous hoplonemerteans*. Am Zool 25:87–97
Stricker SA, Cloney RA (1981) The stylet apparatus of the nemertean *Paranemertes peregrina*: Its ultrastructure and role in prey capture. Zoomorphology 97:205–223
Stricker SA, Cloney RA (1982) Stylet formation in nemerteans. Biol Bull 162:387–403
Tardent P, Holstein T (1982) Morphology and morphodynamics of the stenotele nematocyst of *Hydra attenuata* Pall. (Hydrozoa, Cnidaria). Cell Tissue Res 224:269–290
Thorington GU, Hessinger DA (1988) Control of cnida discharge: factors affecting discharge of cnidae. In: Hessinger DA, Lenhoff HM (eds) The biology of nematocysts. Academic, San Diego, pp 233–253
Thurm U, Brinkmann M, Golz R, Holtmann M, Oliver D, Sieger T (2004) Mechanoreception and synaptic transmission of hydrozoan nematocytes. Hydrobiologia 530:97–105
Tilney LG, Tilney MS (1992) Actin filaments, stereocilia, and hair cells: how cells count and measure. Annu Rev Cell Biol 8:257–274
Verni F (1985) *Litonotus–Euplotes* (predator–prey) interaction: ciliary structure modifications of the prey caused by toxicysts of the predator (Protozoa, Ciliata). Zoomorphology 105:333–335
Watson GM, Hessinger DA (1989) Cnidocyte mechanoreceptors are tuned to the movements of swimming prey by chemoreceptors. Science 243:1589–1591
Watson GM, Hessinger DA (1994) Evidence for calcium channels involved in regulating nematocyst discharge. Comp Biochem Physiol Comp Physiol 107:473–481
Weber J (1990) Poly (γ-glutamic acid)s are the major constituents of nematocysts in *Hydra* (Hydrozoa, Cnidaria). J Biol Chem 265:9664–9669
Wehrmeyer W (1970) Struktur, Entwicklung und Abbau von Trichocysten in *Cryptomonas* und *Hemiselmis (Cryptophyceae)*. Protoplasma 70:295–315
Westfall J, Bradbury PC, Townsend JW (1983) Ultrastructure of the dinoflagellate *Polykrikos*. I. Development of the nematocyst-taeniocyst complex and morphology of the site for extrusion. J Cell Sci 63:245–261
Wood RL, Novak PL (1982) The anchoring of nematocysts and nematocytes in the tentacle of Hydra. J Ultrastruct Res 81:104–116
Zacharias H, Anokhin B, Khalturin K, Bosch TCG (2004) Genome sizes and chromosomes in the basal metazoan *Hydra*. Zoology 107:219–227

Part IV
Applied Evolutionary Biology

Chapter 9
A Possible Relationship Between the Phylogenetic Branch Lengths and the Chaetognath rRNA Paralog Gene Functionalities: Ubiquitous, Tissue-Specific or Pseudogenes

Roxane-Marie Barthélémy(✉), Michel Grino, Pierre Pontarotti,
Jean-Paul Casanova, and Eric Faure

Abstract Chaetognaths constitute a small marine phylum exhibiting two classes of 18S and 28S ribosomal RNA genes which is highly unusual in animal genomes. In situ hybridizations of the chaetognath *Spadella cephaloptera* have shown that 18S class I genes are expressed in the whole body, corresponding to housekeeping genes, whereas 18S and 28S class II genes are expressed in oocytes specifically. Moreover, a heterologous probe against 28S genes revealed only a single signal in a distinct area of intestinal cells. In chaetognaths, the cell gut and oocytes are the cell types which require the greatest translational activities. These results strongly suggest that each type of 18S and 28S paralog is important for specific cellular functions. In addition, phylogenetical analyses have shown a coupling between sequence evolution rate and expression pattern; 28S class II paralogs clearly have the longest branches, indicating a more rapid evolution than class I paralogs.

9.1 Introduction

Chaetognaths are a small marine phylum living in various habitats, but most of them are among the most abundant planktonic organisms (Feigenbaum and Maris 1984). Their body is constituted of three parts, the head, trunk and tail, separated by septa (Casanova 1999). These animals are protandric hermaphrodites; the ovaries lie in

---

R.-M. Barthélémy, P. Pontarotti, J.-P. Casanova, and E. Faure
Laboratoire Evolution Biologique et Modélisation, Case 18, UMR 6632,
Université d'Aix-Marseille/CNRS, 3 Place Victor Hugo, 13331 Marseille CEDEX 03, France
roxane.barthelemy@univ-provence.fr, pierre.pontarotti@univ-provence.fr,
bioplank@up.univ-mrs.fr, eric.faure@univ-provence.fr

M. Grino
Inserm UMR 626, UFR de Médecine Secteur Timone, 27 Bd Jean Moulin,
13385 Marseille CEDEX 5, France
Michel.Grino@medecine.univ-mrs.fr

the trunk on both sides of the gut, while the testes are in the tail. Their phylogenetic position remains enigmatic, although recent molecular analyses suggest a protostome affinity (Shimotori and Goto 2001; Faure and Casanova 2006; Matus et al. 2006; Marletaz et al. 2006). Casanova et al. (2001) showed that chaetognaths can be considered as a model animal.

In eukaryotes, the ribosomal RNA (rRNA) transcription unit usually exists in tandemly repeated arrays containing three rRNA genes, 18S, 5.8S and 28S RNA genes, which are separated by spacer sequences (Prokopowich et al. 2003). To our knowledge, the only animals in which intraindividual variations of the 18S and 28S rRNA genes have been observed are apicomplexans (Rooney 2004), *Acanthamoeba* (Ledee et al. 1998), *Trypanosoma cruzi* (Stothard et al. 1998), platyhelminthes Dugesiidae (Carranza et al. 1999), cephalopods (Bonnaud et al. 2003), lake sturgeon *Acipenser fulvescens* (Krieger and Fuerst 2004; Krieger et al. 2006) and chaetognaths (Telford and Holland 1997; Papillon et al. 2006). Interestingly, a wide distribution of both 18S and 28S classes has been found across the phylum Chaetognatha. Together with the results of phylogenetic analyses (separation between class I and II sequences are supported by high bootstrap values), this fact strongly supports an ancestral duplication of the whole ribosomal gene cluster prior to the radiation of extant chaetognaths, suggesting that the two classes of 18S and 28S rRNA genes are expressed and could be functional (Telford and Holland 1997; Papillon et al. 2006).

9.2 Materials and Methods

Adult specimens of the benthic species *Spadella cephaloptera* were collected during spring 2006 in a marine meadow east of Marseille (Brusc Lagoon, France). The spatial patterns of 18S and 28S paralog expression were examined, using various ribosomal DNA (rDNA) probes, by in situ hybridization on ventral sections (Fig. 9.1; Barthélémy et al. 2007a, b). In situ hybridization with negative control showed, even after a long exposure, no signal pattern, indicating that all the 18S and 28S signals observed are specific.

9.3 Results

9.3.1 18S rRNA Hybridizations

After as little as 4 h of exposure, the X-ray film evidenced that the 18S class I paralog genes are strongly expressed from the head to the tail of the animals, including the ova. When the radioactivity had sufficiently decreased, i.e., 10 weeks after in situ hybridization, it was possible to highlight some differences. The dipped slides show a strong labeling in the gut cells, from the pharynx to the end of intestine, suggesting the strongest expression of the 18S class I genes in this tissue in comparison with the other parts of the body (Fig. 9.1a).

9 Phylogenetic Branch Lengths and the Chaetognath rRNA Paralog Gene Functionalities

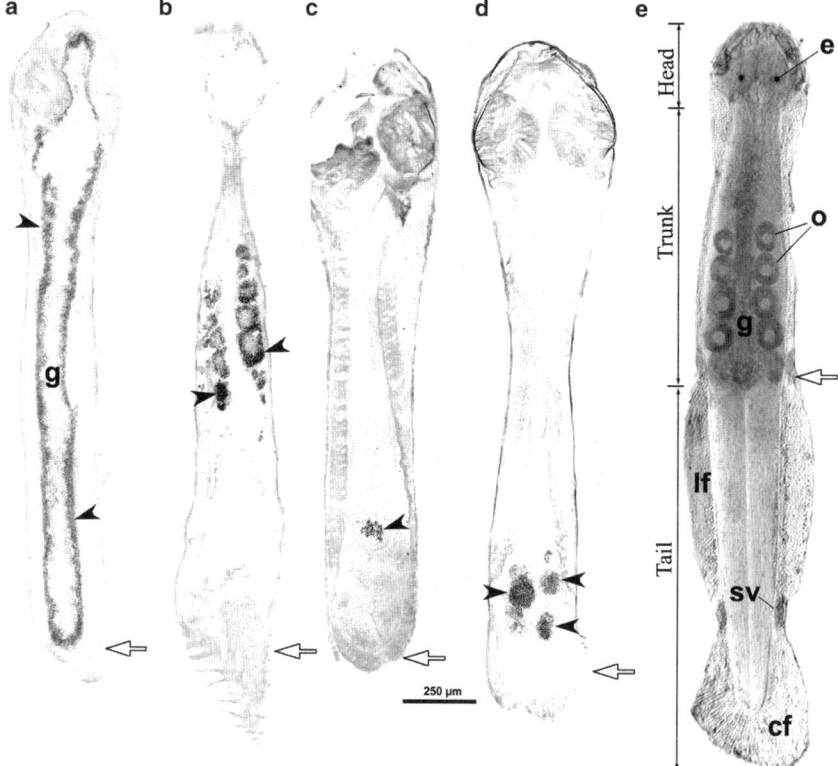

Fig. 9.1 Localization of the 18S and 28S ribosmal DNA paralogs by in situ hybridization (**a–d**, *arrowheads*), using, respectively, antisense 18S class I (**a**), 18S class II (**b**), 28S heterologous class I (**c**) and 28S class II (**d**) genes. Note in **a** the strong expression of the 18S class I genes in the whole gut epithelium, and in **b** that 18S class II genes are expressed only in the cytoplasm of the oocytes; note in **c** weak expression of the 28S class I genes in the gut, and in **d** at the level of oocytes. *Small spots* represent the positive signals, whereas the *gray background* is the result of the coloration process using nuclear fast red. **e** Light microscopy photograph of entire an animal stained with methylene blue with the view to see the anatomy. Note the oocyte cytoplasm around the nucleus as in **b**. *Arrow* transverse septum, *cf* caudal fin, *e* eye, *lf* lateral fin, *g* gut, *o* oocyte, *sv* seminal vesicle

In contrast, the expression of the 18S class II genes began to be visible only after 1 week of exposure. The dipped slides show that this class of genes is specifically expressed in the cytoplasm of the vitellogenic oocytes of individuals during ovogenesis (Fig. 9.1b).

9.3.2 28S rRNA Hybridizations

28S class II genes are strongly and specifically expressed in the oocytes (Fig. 9.1c), whereas use of an heterologous probe against 28S class I genes only resulted in weak expression; indeed, the time needed for visualization of the hybridization signal on the X-ray film was 4 h for the 28S class II probe compared with 6 days with the heterologous class I probe. The rRNAs which hybridize with this last probe probably correspond to a class II variant gene and are located in a restricted area at the end part of the gut (Fig. 9.1d).

9.3.3 Molecular Phylogenies

Two phylogenetic analyses were made, using either 18S or 28S rRNA genes. Twenty-eight sequences of the 28S D2 expansion segment of approximately 500 base pairs from 17 chaetognath species were analyzed. The D2 expansion segment has been found to show relatively high substitution rates in numerous taxa. Similarly to the analyses of both Telford and Holland (1997) and Bompfunewerer et al. (2005), we found that all the 28S class II genes have longer branches than their class I paralogs (Fig. 9.2a).

For 18S analyses, the data set contained 29 18S sequences representing approximately half of the gene size and belonging to 19 chaetognath species (Fig. 9.2b). 18S analyses do not show great differences between class I genes compared with their class II paralogs.

Some rRNA positions have been previously defined as functionally important for small-subunit (SSU) rRNA in prokaryotes (Gutell et al. 1985); moreover, these regions are conserved in some members of the platyhelminthes Dugesiidae, which have two 18S classes. In these animals, the expression of the two rRNA classes has been evidenced (Carranza et al. 1999). To explore further the functionality of chaetognath 18S class I and II genes, the conserved positions were compared with homologous chaetognath regions (Barthélémy et al. 2007a).

This allowed us to identify unambiguously, in the chaetognath 18S rRNA sequences, six positions that possibly act as transfer RNA or messenger RNA (mRNA) contacts; most of the class I and class II chaetognath sequences are similar to the planarian sequences. Moreover, three out of six of these positions are strictly conserved in *Escherichia coli* rRNA sequences (Barthélémy et al. 2007a). Figure 9.2b shows that four sequences have particularly very long branches: two belong to class I (*Sagitta crassa* and *Serratosagitta tasmanica*) and two belong to class II (*Eukrohnia bathypelagica* and *Eukrohnia fowleri*). Moreover, three out of these four 18S chaetognath sequences do not exhibit strict homology sequences with the conserved positions in planarians which have also two rRNA classes, i.e., *E. bathypelagica*, *E. fowleri* and *S. tasmanica*. Moreover, two other sequences belonging to *Krohnitta pacifica* and *Xenokrohnia sorbei* species exhibit shorter branches but without consensus regions (Barthélémy et al. 2007a). In addition, the *S. crassa* 18S class I sequence has some indels (insertion–deletion) which are unique within both chaetognath class I and class II genes (five insertions of one base, one insertion of two bases, one deletion of one base and one deletion of approximately 16 bases near the middle of the sequence). All these results suggest that the longest sequences belonging to both class I and II genes could be pseudogenes.

9 Phylogenetic Branch Lengths and the Chaetognath rRNA Paralog Gene Functionalities

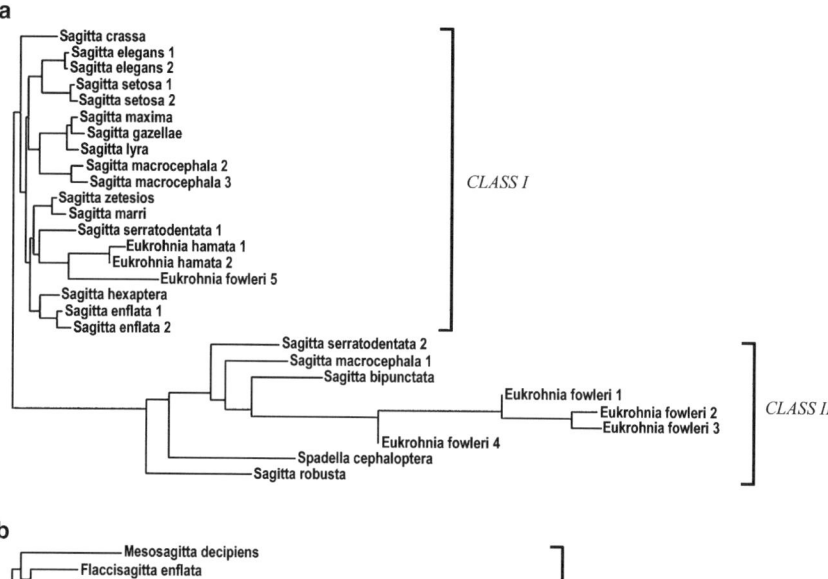

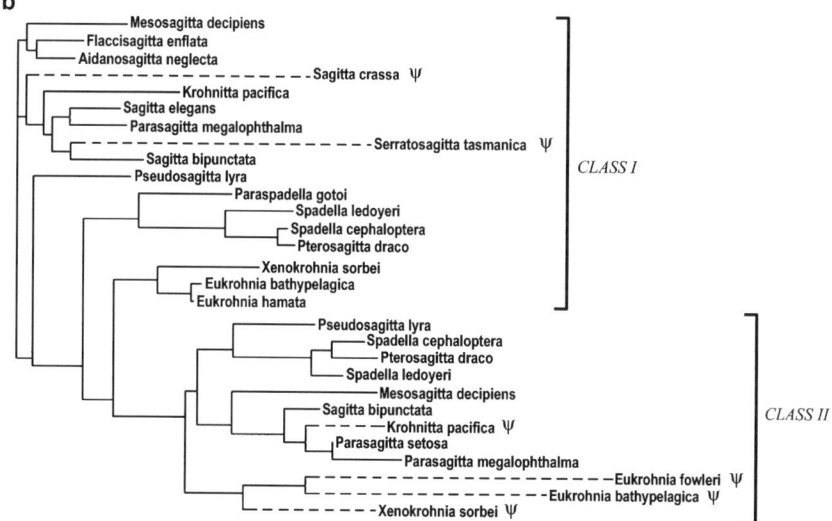

Fig. 9.2 Unrooted neighbor-joining phylogenies of chaetognatha from partial 28S (**a**) and approximately half of the 18S (**b**) rRNA sequences. The trees were recalculated from data reported by Telford and Holland (1997) and Papillon et al. (2006). Both 18S and 28S sequences fall into two paralogous classes that separated at a common ancestor of the recent chaetognaths. In the 28S ribosomal RNA analyses, for the species *Eukrohnia fowleri*, *Sagitta elegans*, *Sagitta macrocephala*, *Sagitta serratodentata* and *Sagitta setosa* several paralogs have been found in the same class. *Sagitta* is synonymous with all the genera containing this word (such as *Flaccisagitta*, *Mesosagitta*, etc.) but in **a** and **b**, these different names have been conserved with the view to unambiguous access in nucleotide sequence databases. In **b**, possible pseudogenes are indicated by *dotted branches* and ψ

9.4 Discussion

For the first time, differential expressions of rRNA paralog genes have been shown in a metazoan, suggesting a ubiquitous role of one of the rRNA classes (class I) and one or more tissue-specific implication(s) for the variant of the other class (Barthélémy et al. 2007a, b). It has been previously shown that in *Plasmodium* species, unicellular parasites (protozoans) that cause malaria, the two 18S variants each have a specific role at different stages of the life cycle (Mercereau-Puijalon et al. 2002).

In chaetognaths, the 18S and 28S class II rRNAs could allow the translation of specific mRNA or maintain a high and continuous level of translation. Contrary to expectations, we have shown that the 28S class I gene products, which hybridize with a heterologous probe, are only expressed in a restricted area at the end part of the gut. As expression of a type of 28S paralog genes in all the parts of the organism during housekeeping conditions is essential for the survival of the animal, our study suggests that at least three types of 28S genes are expressed in the chaetognath *S. cephaloptera*. Probably, the heterologous 28S class I probe allowed the detection of a 28S class II variant.

Interestingly, both 18S and 28S classes II genes and the putative 28S class II variant gene are expressed in tissues where a relatively high level of protein synthesis is necessary. It is known that during oogenesis there is an increase in the levels of materials containing RNA in the cytoplasm, and that *S. cephaloptera* lays a cluster of eggs every day for several days in succession (Ghirardelli 1968). In chaetognaths (Ghirardelli 1968), as in other animal taxa (Canipari et al. 1979), the induction of maturation in oocytes results in an increase in the rate of protein synthesis. Moreover, intestinal cells, either secretory or absorptive, are characterized by an abundant rough endoplasmic reticulum (RER) forming long parallel cisternae covered with many ribosomes. No other somatic tissue of the body (mainly constituted by muscles) exhibits cells with such a high quantity of RER (Shinn 1997). The area shown using the 28S heterologous class I probe corresponds to the central of the three distinct areas of intestinal cells, which still contain dark granules, remaining from digestion, 24 h after feeding. These areas are always present at approximately the same location in all specimens, and the aspect of the cells along the intestine suggests that they correspond to mitotic sites where renewing of intestinal cells take place (Perez 2000; Perez et al. 2000).

These preliminary experiments have shown the presence of paralogous ribosomal protein genes, suggesting that in each protein family some of the members could have tissue-specific and/or extraribosomal functions (Barthélémy et al. 2007c); however, in an alternative hypothesis, each type of ribosomal protein paralog could only interact with one of the rRNA classes, suggesting two very distinct ribosome types. Indeed, 18S in situ hybridizations strongly suggest that a mixed population of the two ribosome types is present in the chaetognath oocytes. A similar event has already been shown during oocyst development in *Plasmodium berghei* (Thompson et al. 1999). In addition, the presence of at least four ribosome populations could not be excluded, i.e., two types of homologous ribosomes (SSU and large subunit, LSU, with either class I or class II rRNAs) and two types of heterologous ribosomes (SSU with 18S class I rRNA and LSU with 28S class II rRNA, and the opposite possibility). Moreover, 28S in situ hybridizations suggest additional possibilities. As the two divergent classes of both 18S and 28S rRNA genes are found in all the extant chaetognaths (Telford and Holland 1997; Papillon et al. 2006), they could have arisen from an allopolyploid event (genome combination after species hybridization) (Barthélémy et al. 2006). The presence of ribosomal protein gene paralogs, which is unusual in animals, is also congruent with this hypothesis (Barthélémy et al. 2007c). Moreover, we can underline that the presence of both rRNA and ribosomal protein paralogs could affect phylogenetic results.

Moreover, the analysis of the cDNA sequences in an expressed sequence tag (EST) library constructed from mRNA isolated from *S. cephaloptera* (Marlétaz et al. 2006) has revealed another putative level of paralogy, complementary DNAs with complete 5' ends encoding a given protein sequence can be divided in two subgroups according to a short region in their 5' ends (Barthélémy et al. 2007c). Three types of these regions have been found in *Sagitta enflata* mRNAs (unpublished data). We have previously hypothesized that the two highly conserved elements found in the 5' ends of mRNAs could correspond to different transcription factor binding sites (Barthélémy et al. 2007c). However, other alternative hypotheses are possible, e.g., each of these two regions could interact with one of the two 18S rRNA classes as ribosomal binding sites. But we cannot also exclude the possibility that some of these regions could be the result of *trans*-splicing maturation on a few mRNAs. Moreover, a great intrapopulational mitochondrial heterogeneity has been found in *S. cephaloptera* ESTs library, suggesting that possible intraindividual heterogeneity must be investigated.

In chaetognaths, 28S class II genes evolve, on average, 1.9 times more rapidly than their class I paralogs; however, owing to the relative small sizes of the sequences which have been analyzed, this value could be overestimated. Concerning 18S analyses, after exclusion of possible pseudogene sequences, the phylogenetic tree does not reveal evident differences between the two paralog classes; however, phylogenies using alignments based on secondary structure prediction should be investigated. Interestingly, in the family Dugesiidae, on average, 18S class II genes evolve 2.3 times more rapidly than class I genes (Carranza et al. 1999); similar results have been found for the lake sturgeon (Krieger and Fuerst 2004). Moreover, for both chaetognaths, dugesiids and sturgeon, the class I genes show more sequence homologies with homologous rRNA genes found in species for which only one class is known than with class II paralogs. Taking into account our hybridization results, analysis of the sequence of the RT PCR product has shown that a single lake sturgeon 18S rDNA sequence variant is expressed in major quantities in the liver; this corresponds to the class I gene which has the shortest branch length (Krieger and Fuerst 2004). In *Dugesia*, northern blot analysis has only shown the transcription of class I genes (Carranza et al. 1996), whereas detection of class II rRNA paralogs has needed RT-PCR detection (Carranza et al. 1999). Taken together, these results strongly suggest that after the duplication (or the allopolyploidy), one of the paralogs has conserved both the ancestral function and the sequence, and has a ubiquitous translational function, whereas class II variants (or some of them) are expressed in low amounts in some minority tissues or play a role during development and regeneration. Moreover, this coupling between sequence evolution rate and expression pattern has been also evidenced by one of us (P.P.) within protein gene families; shorter branch lengths are correlated with ubiquitous tissue expression and tissue expression tends to evolve rapidly for genes expressed in only a limited number of tissues (Balandraud et al. 2005; Paillisson et al. 2007); this is in agreement with our hypothesis. Moreover, the sequence difference level found between the relative rate of class I and of class II genes obtained from the duplication or polyploidy event seems to reinforce the possibility of a subfunctionalization (Lynch and Force 2000; Rastogi and Liberles 2005) which strongly reduces the risk of interference between ubiquitous biological molecules. Owing to this selection pressure, the rate of change of class II sequences has continued to increase long after the duplication or polyploidy event, whereas genes expressed in a large number of tissues can be conserved for a long time. Moreover, multigene families such as rDNA genes have been considered to be special cases of gene duplication since their members are rendered uniform by concerted-evolution mechanisms; in chaetognaths, the presence of two classes of both rRNA and ribosomal protein genes and, to a lesser extent, the great sequence difference levels observed between actin sequences strongly suggest a very low gene conversion level (Barthélémy et al. 2006, 2007c).

9.5 Conclusion

Our results on chaetognaths added to bibliographical data concerning two other metazoan taxa (dugesiids and lake sturgeon) strongly suggest a coupling between sequence evolution rate and expression pattern, i.e., a correlation between the lower sequence evolution rate of class I paralogs (short branch length) and a larger tissue expression distribution, whereas class II paralogs exhibit long branches in relation to a lower or tissue-specific expression level. Similar results have been also found in protein gene families, suggesting this could be a rule in gene families where some of the members have significant sequence differences casued by differences in branch lengths. One of the adaptative advantages of this subfunctionalization is to reduce the risk of deleterious interference with closer functional molecules. Moreover, our results are congruent with the hypotheses of an allopolyploid origin of the chaetognath phylum and of functional ribosome heterogeneity.

References

Balandraud N, Gouret P, Danchin EG, Blanc M, Zinn D, Roudier J, Pontarotti P (2005) A rigorous method for multigenic families functional annotation: the peptidyl arginine deiminase (PADs) proteins family example. BMC Genomics 6:153

Barthélémy R-M, Péténian F, Vannier J, Casanova J-P., Faure E (2006) Evolutionary history of the chaetognaths inferred from actin and 18S-28S rRNA paralogous genes. Int J Zool Res 2:284–300

Barthélémy R-M, Grino M, Pontarotti P, Casanova J-P, Faure E (2007a) The differential expression of ribosomal 18S RNA paralog genes from the chaetognath *Spadella cephaloptera*. Cell Mol Biol Lett 12:573–583

Barthélémy R-M, Casanova J-P, Grino M, Faure E (2007b) Selective expression of two types of 28S rRNA paralogous genes in the chaetognath *Spadella cephaloptera*. Cell Mol Biol (Noisy-le-grand) 53(Suppl):OL989–993

Barthélémy R-M, Chenuil A, Brancart S, Casanova J.-P, Faure E (2007c) Translational machinery of the chaetognath *Spadella cephaloptera*: a transcriptomic approach to the analysis of cytosolic ribosomal protein genes and their expression. BMC Evol Biol 7:146

Bompfunewerer AF, Flamm C, Fried C, Fritzsch G, Hofacker IL, Lehmann J, Missal K, Mosig A, Muller B, Prohaska SJ, Stadler BMR, Stadler PF, Tanzer A, Washietl S, Witwer C (2005) Evolutionary patterns of non-coding RNAs. Theor Biosci 123:301–369

Bonnaud L, Saihi A, Boucher-Rodoni R (2003) Are 28SrDNA and 18SrDNA informative for cephalopod phylogeny? Bull Mar 71:197–208

Canipari R, Pietrolucci A, Mangia F (1979) Increase of total protein synthesis during mouse oocyte growth. J Reprod Fertil 57:405–413

Carranza S, Giribet G, Ribera C, Baguñà, Riutort M (1996) Evidence that two types of 18S rDNA coexist in the genome of *Dugesia* (Schmidtea) mediterranea (Platyhelminthes, Turbellaria, Tricladida). Mol Biol Evol 13:824–832

Carranza S, Baguna J, Riutort M (1999) Origin and evolution of paralogous rRNA gene clusters within the flatworm family Dugesiidae (Platyhelminthes, Tricladida). J Mol Evol 49:250–259

Casanova J-P (1999) Chaetognatha. In: Boltovskoy D (ed) South Atlantic zooplankton. Backhuys, Leiden, pp 1353–1374

Casanova J-P, Duvert M, Perez Y (2001) Phylogenetic interest of the chaetognath model. Mésogée 59:27–31.

Faure E, Casanova J-P (2006) Comparison of chaetognath mitochondrial genomes and phylogenetical implications. Mitochondrion 6:258–262

Feigenbaum DL, Maris RC (1984) Feeding in chaetognatha. Oceanogr Mar Biol Annu Rev 22:343–392

Ghirardelli E (1968) Some aspects of the biology of the Chaetognaths. Adv Mar Biol 6:271–375

Gutell RR, Weibser B, Woese CR, Noller HF (1985) Comparative anatomy of 16S-like ribosomal RNA. Prog Nucleic Acid Res Mol Biol 32:155–216

Krieger J, Fuerst PA (2004) Characterization of nuclear 18S rRNA gene sequence diversity and expression in an individual lake sturgeon (*Acipenser fulvescens*). J Appl Ichthyol 20:433–439

Krieger J, Hett AK, Fuerst PA, Birstein VJ, Ludwig A (2006) Unusual intraindividual variation of the nuclear 18S rRNA gene is widespread within the acipenseridae. J Hered 97:218–225

Ledee DR, Seal DV, Byers TJ (1998) Confirmatory evidence from 18S rRNA gene analysis for in vivo development of propamidine resistance in a temporal series of *Acanthamoeba* isolates from a patient. Antimicrob Agents Chemother 42:2144–2145

Lynch M, Force A (2000) The probability of duplicate gene preservation by subfunctionalization. Genetics 154:459–473

Marletaz F, Martin E, Perez Y, Papillon D, Caubit X, Lowe CJ, Freeman B, Fasano L, Dossat C, Wincker P, Weissenbach J, Le Parco Y (2006) Chaetognath phylogenomics: a protostome with deuterostome-like development. Curr Biol 16:R577–578

Matus DQ, Copley RR, Dunn CW, Hejnol A, Eccleston H, Halanych KM, Martindale MQ, Telford MJ (2006) Broad taxon and gene sampling indicate that chaetognaths are protostomes. Curr Biol 16:R575–576

Mercereau-Puijalon O, Barale JC, Bischoff E (2002) Three multigene families in *Plasmodium* parasites: facts and questions. Int J Parasitol 32:1323–1344

Paillisson A, Levasseur A, Gouret P, Callebaut I, Bontoux M, Pontarotti P, Monget P (2006) Bromodomain testis-specific protein is expressed in mouse oocyte and evolves faster than its ubiquitously expressed paralogs BRD2, -3, and -4. Genomics 89:215–223

Papillon D, Perez Y, Caubit X, Le Parco Y (2006) Systematics of Chaetognatha under the light of molecular data, using duplicated ribosomal 18S DNA sequences. Mol Phylogenet Evol 38:621–634

Perez Y (2000) L'appareil digestif des chaetognathes, structure et ultrastructure, aspects fonctionnels et écophysiologiques. Doctoral thesis, Université de Provence, Marseille

Perez Y, Casanova J-P, Mazza J (2000) Changes in the structure and ultrastructure of the intestine of *Spadella cephaloptera* (Chaetognatha) during feeding and starvation experiments. J Exp Mar Biol Ecol 253:1–15

Prokopowich CD, Gregory TR, Crease TJ (2003) The correlation between rDNA copy number and genome size in eukaryotes. Genome 46:48–50

Rastogi S, Liberles DA (2005) Subfunctionalization of duplicated genes as a transition state to neofunctionalization. BMC Evol Biol 5:28

Rooney AP (2004) Mechanisms underlying the evolution and maintenance of functionally heterogeneous 18S rRNA genes in apicomplexans. Mol Biol Evol 21:1704–1711

Shimotori, T, Goto T (2001) Developmental fates of the first four blastomeres of the chaetognath *Paraspadella gotoi*: relationship to protostomes. Dev Growth Differ 43:371–382

Shinn GL (1997) Chaetognaths. In: Harrison FW, Ruppert EE (eds) Microscopic anatomy of invertebrates, vol 15. Hemichordates, Chaetognatha and the invertebrate chordates. Wiley-Liss, New York, pp 103–220

Stothard JR, Frame IA, Carrasco HJ, Miles MA (1998) Temperature gradient gel electrophoresis (TGGE) analysis of riboprints from *Trypanosoma cruzi*. Parasitology 117:249–253

Telford MJ, Holland PWH (1997) Evolution of 28S ribosomal DNA in Chaetognaths: duplicate genes and molecular phylogeny. J Mol Evol 44:135–144

Thompson J, van Spaendonk RM, Choudhuri R, Sinden RE, Janse CJ, Waters AP (1999) Heterogeneous ribosome populations are present in *Plasmodium berghei* during development in its vector. Mol Microbiol 31:253–360

Chapter 10
Mode and Tempo of *matK*: Gene Evolution and Phylogenetic Implications

Khidir W. Hilu and Michelle M. Barthet

Abstract The *matK* gene is approximately 1,500 base pairs in length, located in the large single copy region of the chloroplast genome, nested in the *trnK* intron. The protein product of the gene is proposed to function as a group II intron maturase. Evolutionary, molecular, and bioinformatic evidence supports this proposed function. The open reading frame of this protein-coding gene, however, has a high proportion of substitutions at the nucleotide and amino acid levels and has accumulated substantial numbers of insertions/deletions. These features have rendered *matK* as a valuable source of phylogenetic signal in resolving relationships in plants at various historic levels but have also raised questions about the functionality of the gene. We discuss in this chapter how the unique combination of high substitution rates and mode of evolution that approach neutrality results in high phylogenetic signal and preservation of protein function in *matK*.

10.1 Introduction

The *matK* gene is located in the large single copy region of the chloroplast genome, in the vicinity of one of the inverted repeat regions (Fig. 10.1). The open reading frame (ORF) of the gene is about 1,500 base pairs (bp) in length in most plants. With two known exceptions, this ORF is nested in the *trnK* intron that codes for transfer RNA-lysine$^{(UUU)}$. The two exceptions are cases where the *trnK* intron was lost owing to an inversion in the plastid genome of the maidenhair fern *Adiantum*

K.W. Hilu
Department of Biological Sciences, Virginia Tech, Blacksburg, VA 24061, USA
hilukw@vt.edu

M.M. Barthet
ARC Centre of Excellence in Plant Energy Biology, School of Biological Sciences, University of Sydney, Sydney, NSW 2006, Australia
michelle.barthet@bio.usyd.edu.au

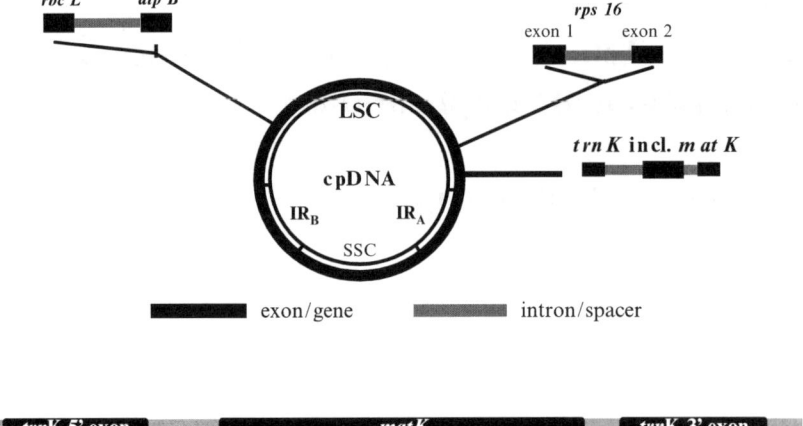

Fig. 10.1 The location of *trnK/matK* in the chloroplast genome. *Top*: Relative positions of the inverted repeats (*IR*) and the large single copy region (*LSC*) of the chloroplast genome of higher plants along with the positions of a few other genes commonly used in plant systematics. *trnK/matK* is highlighted. *Bottom*: The relative position of *matK* to the 5′ and 3′ *trnK* exons

capillus-veneris (Wolf et al. 2003), or as one of many plastid genes was lost in the holoparasite *Epifagus virginiana* (Ems et al. 1995; Wolfe et al. 1992). The 3′ end of the *matK* gene contains a highly conserved region with homology to domain X, the maturase domain of mitochondrial group II intron maturases (enzymes that catalyze intron removal from premature RNAs) (Neuhaus and Link 1987). *matK* is the only chloroplast gene containing this type of enzymatic domain (du Jardin et al. 1994), making it the only putative group II intron maturase encoded in the chloroplast. On the basis of the conservation of this domain, *matK* was hypothesized to function in the splicing of the group II intron in which it is nested (Neuhaus and Link 1987; Ems et al. 1995), and possibly other group II introns (Ems et al. 1995; Vogel et al. 1997, 1999).

The *matK* gene exhibits unusual mode and tempo of evolution across land plants compared with plastid genes that have been examined in an evolutionary context. The gene exhibits relatively high proportions of substitutions across most of its ORF, undergoes unusually high amounts of nonsynonymous mutations for a protein-coding gene, and is laden with insertion/deletion events, or indels (Olmstead and Palmer 1994; Johnson and Soltis 1994; Hilu and Liang 1997; Soltis and Soltis 1998; Hilu et al. 2003; Müller et al. 2006). The gene also exhibits a relatively high proportion of transversions, with the transition to transversion ratio approaching unity (Olmstead and Palmer 1994; Hilu and Liang 1997). The rate of substitution per site in *matK* is threefold that found in the plastid gene ribulose 1,5-bisphosphate carboxylase (*rbcL*) (Hilu and Liang 1997; Xiang et al. 1998). The almost equal rates of substitution among the three codon positions (Hilu and Liang 1997; Hilu et al. 1999) are in sharp contrast with the skewed rates of substitution towards the third codon position found in most protein-coding genes. Subsequently, the percentage of amino acid substitution is considerably elevated in *matK*. For instance, amino acid substitution for *matK* between the monocot rice and the eudicot tobacco is up to six-fold higher than for *rbcL* and *atpB* (41 vs. 7–8%; Olmstead and Palmer 1994). The unusually high rate of overall substitution at the nucleotide level, the elevated rates of amino acid substitution, and the prevalence of gaps in the ORF have not only placed *matK* in a different category in terms of gene mode and tempo of evolution, but have also raised questions about its functionality. Traditionally, it is an accepted notion that excessive changes in amino acids and loss or gain of nucleotide motifs should impact gene structure and consequently function, and thus are indications of potential loss of functionality. Consequently, the question was raised of whether *matK* is a functional gene or a mere pseudogene. The implications of the functionality (or lack of it) of *matK* are far-reaching in terms of physiological function of the plastid organelle since *matK* is the only putative group II intron maturase encoded in the chloroplast. This issue will be addressed later in this chapter.

The *matK* gene has also become a valuable source of molecular characters for the assessment of plant diversity on a wide historic scale, from the genus level to as deep as land plant diversity (Hilu et al. 2003). The wealth of mutational events it possesses, both substitutions and indels of DNA stretches, provides a strong phylogenetic signal that surpasses that of most genomic regions used in plant molecular phylogenetics. The questions concerning the increased utility of sequences from this gene in plant phylogenetics are twofold. One, why is *matK* so astonishingly informative to the point that information from only two thirds of the gene provides a phylogentic structure for flowering plants that is as good as sequences from two to 11 genes combined (Hilu et al. 2003)? Two, if *matK* is a nonfunctional gene, how useful is it for resolving issues in plant phylogenetics? Both of these points will be addressed here.

10.2 Gene Structure

The approximately 1,500 bp *matK* ORF is located in the large single copy region of the chloroplast genome nested within the group II intron between the 5′ and 3′ *trnK*$^{(UUU)}$ exons in most green plants (Sugita et al. 1985; Steane 2005; Daniell et al. 2006; Fig. 10.1). Nucleotide substitution rates in this gene are up to three-times that of *rbcL* (Hilu and Liang 1997; Xiang et al. 1998). These substitutions are distributed almost equally among all three codon positions. For example, the substitution rates are 62, 57, and 66%, for codon positions 1, 2, and 3, respectively, when comparing at the ordinal level across angiosperms (Hilu et al. 2003). A similar pattern of substitution was observed in studies at the family level, e.g., 32, 28, and 39%, respectively, for members of Orchidaceae (Whitten et al. 2000). The even distribution of substitutions has resulted in a nonsynonymous substitution rate sixfold that of *rbcL* and other chloroplast genes (Olmstead and Palmer 1994). Indels are also commonly found within the *matK* ORF. These indels range in size and number depending on the species being compared, but are primarily found in multiples of three conserving the reading frame (Hilu and Liang 1997; Hilu et al. 2003; Quint and Classen-Bockhoff 2006). For example, *Epifagus matK* has a 204-bp deletion starting at the first position of the 5′ end of the gene when comparing across 11 land plants, while other plants used in this same study displayed up to 17 smaller indels of 3–15 bp (Hilu and Liang 1997). Four indels in multiples of three were found when aligning *matK* sequences of the Bruniaceae, and three indels were identified for just the 3′ region of the gene in the Poaceae (Hilu and Alice 1999; Quint and Classen-Bockhoff 2006). Further, transition to transversion ratios approach unity when examining sequences at deep (ordinal) or shallow (subfamily) taxonomic levels (Liang and Hilu 1996; Hilu et al. 2003). Although the nucleotide and amino acid sequences from this gene reveal a pattern of high mutation, these substitutions are constrained by the amino acid side chain category (i.e., acid, base, polar, nonpolar, etc.; Fig. 10.2) and in predicted secondary structure across green plants (Barthet and Hilu 2008).

The *matK* gene can be divided into two general regions, the N-terminal region and domain X (Fig. 10.3). The N-terminal region contains remnants of the reverse-transcriptase (RT) domain common to other group II intron maturases (Mohr et al.

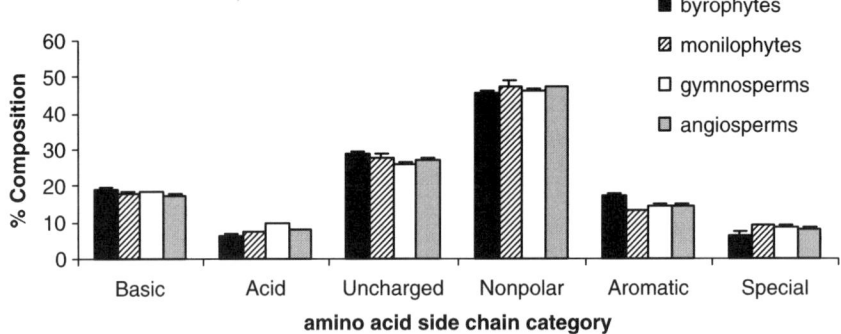

Fig. 10.2 Composition of amino acid side chain categories in MATK in four major lineages of land plants. Amino acids were categorized into the six groups: basic (K, R, H), acid (D, E), amino acids, pH 7 (N, Q, S, T, Y), nonpolar (A, G, V, L, I, P, F, M, W, C), aromatic (F, W, Y), and "special" (amino acids that strongly impact protein structure; C, G, P) following Barthet and Hilu (2008). The taxa used were bryophytes (*Physcomitrella*, *Marchantia*, *Huperzia*, and *Anthoceros*), monilophytes (*Psilotum* and *Adiantum*), gymnosperms (*Pinus*, *Cryptomeria*, *Phyllocladus*, *Taxus*, and *Sequiadendron*), and angiosperms (*Amborella*, *Nypa*, *Oryza*, *Nelumbo*, *Platanus*, and *Spinaceae*). The species are those used in Barthet and Hilu (2008)

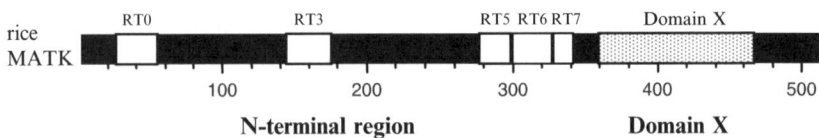

Fig. 10.3 The structural elements found within MATK. Positions of sequence blocks of the reverse transcriptase (RT) domain and domain X in MATK were determined by homology in secondary structure (Barthet and Hilu 2008) or sequence (Mohr et al. 1993)

1993; Barthet and Hilu 2008), while domain X is considered the site of maturase activity (Cui et al. 2004). Studies have noted low nucleotide variability and a lack of indels within the region denoted domain X in the 3' end of *matK* (Hilu and Liang 1997; Hilu et al. 2003), with much less constraint observed in the 5' N-terminal region (Hilu and Liang 1997). More recent studies, however, indicated relatively equal levels of constraint between the N-terminal region and domain X (Young and dePamphilis 2000; Barthet and Hilu 2008). Analysis of variation of the amino acid chemical composition indicated that there are at least three highly constrained regions in MATK (the protein product of the *matK* gene), two of which reside within the N-terminal region (Barthet and Hilu 2008). Further, secondary structure alignment has shown that several structural elements of the RT domain found in other group II intron maturases, stretching from the very beginning of the N-terminal region to domain X, are conserved in *matK* (Barthet and Hilu 2008; Fig. 10.3). The relationship between conserved elements of the gene structure and the putative function of MATK as a group II intron maturase will be discussed in Sect. 10.6.

10.3 Evolution of *matK*

Phylogenetic analysis of the RT domain found in most intron-encoded proteins (IEPs), such as MATK, revealed high similarity to the RT domains of non-long-terminal-repeat retroelements (non-LTRs) (Mohr et al. 1993). From this similarity it was proposed that these non-LTRs are the evolutionary ancestors of IEPs, including MATK, and this is referred to as the "retroelement ancestor hypothesis" (Hausner et al. 2006). Non-LTRs are also thought to be the ancestors of retroviral RTs like HIV-1 (Eickbush and Malik 2002). Examination of the group II intron of *trnK* among different plant lineages revealed that this intron, presumably with the *matK* ORF, integrated between the *trnK* exons sometime after the divergence of chlorophytes and *Mesostigma* from the algal lineage, approximately 1,200–800 million years ago (Hausner et al. 2006). The general structure of group II intron maturases includes three parts: a complete RT domain for retrotransposon mobility, the X domain, and a zinc-finger-like region at the C-terminal end (Mohr et al. 1993; Hausner et al. 2006; Fig. 10.3). Unlike many other IEPs, *matK* lacks a complete RT domain and, thus, has lost mobility. From this point in evolution, the *trnK* intron, including the *matK* ORF, could only be inherited vertically (Hausner et al. 2006).

10.4 *matK* Role in Plant Phylogenetics

In the past 15 years, there has been a considerable leap in understanding patterns of biological diversity. This immense progress can be attributed in large to accelerated advances in the fields of molecular biology, particularly DNA sequencing technology, and bioinformatics, at both software and hardware levels. In plants, numerous traditional evolutionary and systematic concepts have been revised, and in some cases changes were dramatic, such as in flowering plants or angiosperms (Savolainan et al. 2000; APG II, 2003; Hilu et al. 2003) and lower vascular plants (Pryer et al. 2001; Qiu et al. 2006). We will focus here on the changes that impacted flowering plants, the most diversified and largest group of land plants.

In flowering plants (angiosperms), the concepts of the basal split into monocots and dicots, the proposed most basal position of a magnolia-like plant in the angiosperm tree, and the division of angiosperms into 11 subclasses *sensu* Cronquist (1988) has been challenged with evidence from molecular phylogenies and cladistic analyses of morphological and structural characters (Nandi 1998; Mathews and Donoghue 2000; Savolainen et al. 2000; Zanis et al. 2002; APG II 2003; Hilu et al. 2003). These studies clearly demonstrated the absence of a basal split into monocots and dicots, but instead showed monocots to be nested deeper in the angiosperm tree. The current consensus is that the New Caledonian shrub *Amborella* (Amborellaceae) occupies the most basal position of the angiosperm tree, but not a Magnoliaceae plant, and that the 11 subclasses circumscribed by Cronquist are not a valid taxonomic grouping of the angiosperm diversity. Currently, in the flowering plants tree of life a basal grade of three consecutive lineages of Amborellaceae, Nymphaeales, and Austrobaileyales is almost always resolved, followed by the three large lineages of magnoliids (containing the Magnoliaceae), monocots, and eudicots, plus the two families Ceratophyllaceae (Ceratophyllales) and Chloranthaceae (Chloranthales) (Fig 10.4). The relationships among the latter five lineages remain inconclusive.

Available fossil records estimate the age of flowering plants at approximately 130 million years. The assessment of the phylogenetic relationships at this relatively deep historic level has been done primarily with slowly evolving genes. This approach is based on the premise that slowly evolving genes are more suitable for reconstructing phylogenies over long periods of evolutionary time, as their individual nucleotide sites are less likely to undergo multiple hits (Farris 1977; Swofford et al. 1996; Graham and Olmstead 2000). Multiple hits, predicted to be prevalent in rapidly evolving genes, could conceal signal and possibly provide noise (homoplasy) that may negatively impact phylogenetic reconstruction. As a consequence of this concept, the first gene to be used in a large-scale phylogenetic analysis of angiosperm was *rbcL* (Chase et al. 1993). This study was soon followed by the addition of sequence information from yet two more slowly evolving genes, the plastid gene *atpB* (Savolainen et al. 2000) and then the nuclear 18S ribosomal RNA gene (Soltis et al. 2000) to generate a two and three combined genes data set. The addition of sequences from these two genomic regions resulted in some changes in the tree topologies and added more resolution and support for numerous relationships. The concept of the effectiveness of a slowly evolving gene was taken to a new level with the study of Graham and Olmstead (2000) where sequences from 17 slowly evolving plastid genes, amounting to 13.4 kilobase characters, were combined to assess the impact of increasing molecular characters from slowly evolving genes on phylogenetic reconstruction in early diverging angiosperms. The resolution and support obtained from this latter study was not as satisfactory, particularly when one considers the time, effort, and expense incurred from using such a large number of

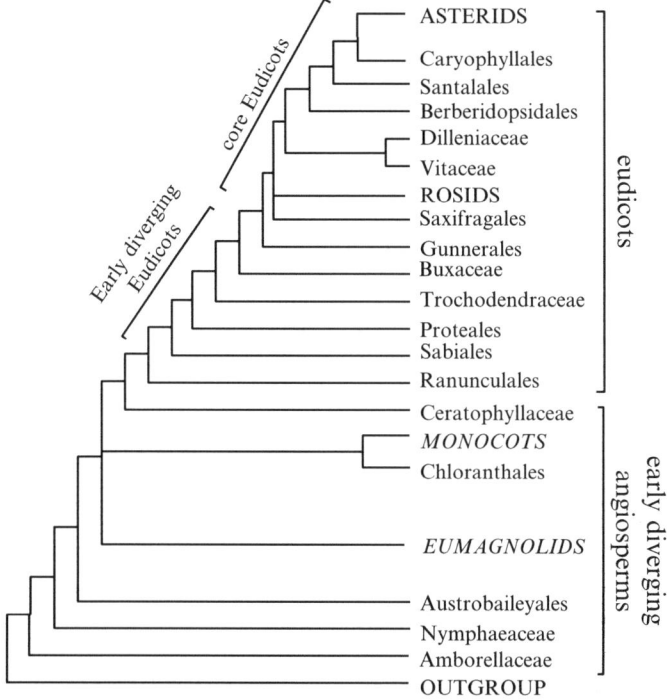

Fig. 10.4 Angiosperm phylogeny resolved from phylogenetic analysis of *matK* sequences (1,200 base pairs). The tree depicts Amborellaceae as sister to remaining angiosperms and highlights the major flowering plant lineages. *Uppercase letters* signify large-sized groups and *italics* refer to the new internal position of the Magnoliaceae and the monocots in the flowering plants tree compared with traditional concepts of a basal position (see the text). (Modified from Hilu et al. 2003)

genes. In addition, the use of such a large number of sequences adds a constraint on the number of species that can be sampled, which by itself can introduce errors in phylogenies.

10.5 Why Is *matK* So Signal Rich?

Using the plastid gene *matK*, we proceeded in three studies (Hilu and Liang 1997; Hilu et al. 2003; Müller et al. 2006) to examine the concept that rapidly evolving genes are not an appropriate measure of deep-level phylogenetic relationships. In our first study (Hilu and Liang 1997), we examined the distribution of variability and phylogenetic signal horizontally in the *matK* ORF, and vertically across seed plants from between two species of pines to deep into the base of the tree. The results of the study showed the *matK* ORF to be for the most part uniform in terms of variability and that mutational events across the gene are useful in discerning relationships among species as well as deep in the phylogenetic tree of the species included. The resolution obtained for the tree from that study, despite the relatively small sample size, was quite impressive for a single gene data set. The strong test of the utility of *matK* at deep-level phylogentics was established in a collaborative study using a large data set of 374 genera representing all angiosperm orders and 12 gymnosperm genera (Hilu et al. 2003). Although the data set included only two thirds of the gene, since a good number of the sequences lacked about 400 bp from the 5' section of the gene, the resolution and support obtained by partial *matK* was quite impressive. The relationships discerned from the partial sequences of the *matK* gene matched those that used sequences from three to 11 genes combined.

The question is, why is *matK* so informative in phylogenetic reconstruction particularly when compared with most genes used in similar studies? Is it because of the large number of variable sites or the signal per site in *matK* that renders the gene to be superior to other genes used in deep-level phylogenetics. To address this question, we conducted a study in which we used sequences of the slowly evolving *rbcL*, the rapidly evolving *trnL-trnF* spacer regions, and *matK* (all plastid genomic regions) from the same 42 species representing mostly early diverging flowering plant lineages (Müller et al. 2006). *MatK* provided about 3 times as many parsimony informative sites as *rbcL* (287 vs. 862), which correspond to 24 vs. 52% of the variable characters found in the two genomic regions. The phylogenetic trees based on the individual genomic region *matK* and *trnL-trnF* sequences provided far higher resolution and support compared with the one based on *rbcL* (Müller et al. 2006). To see if there is a difference in the magnitude of the phylogenetic signals (quality of signal) inherent to these genomic regions, we conducted a resampling approach with an equal number of parsimony informative sites and the results were evaluated statistically for overall tree robustness. Our results showed that *matK* not only provided more parsimony informative characters, but also that the information per site is statistically stronger than that obtained from *rbcL*. The relative performance of *matK* correlates not only to a lower mean homoplasy in *matK* compared with *rbcL*, but also to a negative correlation with the percentage of sites exhibiting maximum or close to maximum homoplasy. The study concluded that the negative impact of multiple hits in using rapidly evolving DNA at relatively deep phylogenetic levels may have been overestimated, and thus the use of rapidly evolving DNA should be extended to deeper phylogenetic levels. This finding is in line with the conclusion from the study of Källersjö et al. (1999) that homoplasy does not necessarily negatively impact tree robustness.

These notable quantitative and qualitative features of the signal richness of *matK* logically lead to another question, namely, what are the inherent molecular evolutionary drives in *matK* that render it so phylogenetically informative? One of the answers came from our recent study (Müller et al. 2006) comparing distances at synonymous sites with those at nonsynonymous sites in *matK* and *rbcL*. The *rbcL* gene deviates more significantly from neutrality than *matK*. Evidence for the greater degree of purifying selection displayed in *rbcL* compared with *matK* can be extrapolated from comparing the number of steps per codon position in trees based on the two genes. The number of steps in *rbcL* for third positions were 10 times those of second positions, contrasted with approximately 1.5 for *matK*. These results point to differences in degrees of purifying selection operating on these two genes when based on the theoretical frequencies of synonymous and nonsynonymous mutations at the three codon positions (Li and Graur 1991). These findings are intriguing considering the view that synonymous substitutions are regarded as neutral or silent mutations whereas nonsynonymous substitutions are indicative of selective pressures (Ophir et al. 1999; Young and dePamphilis 2005).

Further evidence for the neutrality of *matK* was provided by the study of Barthet and Hilu (2008). In that study, we compared variation in side chain composition of MATK (the protein product of the *matK* gene) with that of RBCL, INFA, and MATR in order to establish the mode and relative degree of evolutionary constraint on MATK. INFA is a plastid pseudogene and MATR is a mitochondrial maturase. Variation in side chain composition of MATK is higher than that of the conserved protein RBCL but not as fast as that of the pseudogene INFA; therefore, it appears that mutations in MATK are in general neutral and do not result in change (positive or negative) in protein structure. This notion gains support from the finding in the same study that nonsynonymous mutations in MATK tend to be constrained within amino acid chemical groups, i.e., from acidic to acidic or from basic to basic chemical classes, conserving the structure and function of the gene. Barthet and Hilu (2008) proposed that most of these substitutions could be construed as silent mutations, underscoring the neutral nature of *matK* mutations. Therefore, it seems that a high number of variable sites, higher proportions of phylogenetically informative sites, and the tendencies toward neutrality combined result in the ability of *matK* to provide this robustness in phylogenetic reconstruction.

10.6 Support for Function

Sugita et al. (1985) first identified the *matK* gene, originally noted as ORF509, within the group II intron of *trnK* in the chloroplast genome of tobacco in 1985. Two years later, Neuhaus and Link (1987) found a homologous ORF within the chloroplast *trnK* group II intron of the mustard *Sinapsis alba*. Neuhaus and Link (1987) further determined that the *trnK* intron ORF (*matK*) from both *S. alba* and tobacco contained a region near its carboxy terminus with high homology to domain X of group II intron maturases. The identification of this domain led to the proposition that *matK* protein functions as a group II intron maturase in the chloroplast.

Introns can be broadly grouped as group I or group II introns (reviewed in Saldanha et al. 1993). Group I introns are autocatalytic, using only an internal guanosine residue and Mg^{2+} for splicing (reviewed in Saldanha et al. 1993). In contrast, group II introns are generally nonautocatalytic, requiring additional splicing factors (e.g., a maturase) for intron removal (reviewed in Saldanha et al. 1993; Noah and Lambowitz 2003). There are 16 group II introns nested within transcripts of the chloroplast genome of higher plants (Vogel et al. 1999, 1997) which require a maturase for intron splicing and proper protein translation. Some of these transcripts have been shown to need a plastid-encoded factor for maturation (Vogel et al. 1999, 1997). Since *matK* is the only gene encoded in the chloroplast genome containing the maturase domain X (du Jardin et al. 1994), it may be a critical enzyme for chloroplast function. The function of MATK as a group II intron maturase has been supported by evolutionary, bioinformatics, and molecular analyses.

10.6.1 Evolutionary Evidence

The location of the *matK* gene throughout most of plant evolution has been within the group II intron of *trnK*. There are, however, two known exceptions: (1) in the maiden-hair fern *Adiantum* and (2) in the holoparasitic plant *Epifagus virginiana*. The restructuring of the chloroplast genomes in these two species provides strong evolutionary evidence for the importance of *matK* as a functional gene in plants. The loss of *trnK* from the chloroplast genome of *Adiantum* occurred through a large rearrangement in the plastid genome. This rearrangement appears to have occurred in the *trnK* intron and resulted in the loss of *trnK* but retention of *matK* (Wolf et al. 2003). In spite of the loss of *trnK*, a *matK* transcript is still evident in *Adiantum capillus-veneris* (Wolf et al. 2004), supporting independent expression of this gene.

Eleusine virginiana lost over 60% of the genes in its chloroplast genome, including *trnK* (Ems et al. 1995). The *matK* gene, though, was retained in this residual chloroplast genome. The persistence of *matK* along with the 40% of genes that are needed for the survival of this plant implies that MATK must serve an essential function in the plant. Of the 42 genes that are maintained in the residual chloroplast genome of *E. virginiana*, six of these contain group II introns, which would require a maturase for splicing (Ems et al. 1995). Five of these six were shown to be spliced properly in *E. virginiana*, and two of these five, *rpl2* (ribosomal protein L2) and intron 2 of *rps12* (ribosomal protein S12), were shown in biochemical studies to require a plastid-encoded factor (MATK?) for intron processing (Ems et al. 1995; Hess et al. 1994; Hübschmann et al. 1996). From these results, it was hypothesized that the *matK* gene was retained in the *E. virginiana* residual genome to process group II introns from these transcripts (Ems et al. 1995). In both cases, that of *Adiantum* and that of *E. virginiana*, the *trnK* group II intron was lost from the chloroplast genome, and thus MATK would not be required to process the *trnK* intron in these plants. The continued existence of *matK* in these genomes and the documentation of its transcription in *Adiantum* (Wolf et al. 2004) imply that MATK could potentially splice group II introns other than that of *trnK*.

10.6.2 Evidence from Bioinformatics

Support for MATK function has also been found using bioinformatics methods, specifically, through alignments of predicted *matK* amino acid sequence and protein structure. These alignments have shown that MATK contains all the essential elements for maturase activity. Most maturases contain three conserved functional domains, a RT domain, domain X, and a C-terminal zinc-finger matK-like region (Mohr et al. 1993). Domain X is considered the primary domain required for maturase activity. Key mutations in domain X of the bacterial group II intron maturase LTRA were shown to completely inhibit maturase activity (Cui et al. 2004). In addition, mutations in elements of the RT domain also result in reduced maturase activity, also implicating the RT domain as important for efficient intron splicing (Cui et al. 2004). Initial alignment of the predicted amino acid sequence of MATK to other maturases displayed high homology to domain X of other group II intron maturases, as well as to three conserved sequence blocks of the RT domain, sequence blocks 5, 6, and 7 (Mohr et al. 1993). A recent alignment using secondary structure showed structural homology in MATK to these three as well as two other (RT0 and RT3) sequence elements in the RT domain (Barthet and Hilu 2008; Fig. 10.3). The intact X domain of MATK suggests that this protein remains capable of RNA splicing. The conserved elements of the RT domain found in MATK are most likely required for binding of the template RNA transcript for more efficient maturase activity. The remaining elements of the RT domain that are not maintained in MATK (sequence blocks 1, 2, and 4) have been implicated as some of the factors involved in determining intron specificity for activity (Cui et al 2004). The loss of these elements from the *matK* ORF suggests that this protein is still able to bind substrate and process introns but lacks specificity. This would enable MATK to splice several intron substrates. Most maturases are presumed to only process the group II intron in which they are located (Saldanha et al. 1999; Moran et al. 1994; Rambo and Doudna 2004). It was hypothesized, however, on the basis of evolutionary evidence from *E. virginiana* and *Adiantum* as well as biochemical assays, which will be discussed next, that MATK has the ability to splice several group II intron containing transcripts.

10.6.3 Molecular Evidence

A protein product for *matK* has been observed in protein extracts from potato (du Jardin et al. 1994), mustard (Liere and Link 1995), barley (Vogel et al. 1999), rice, arrowhead, sugarcane, oat, and arabidopsis (Barthet and Hilu 2007). The molecular mass of this protein ranges from 43 kDa in potato (du Jardin et al. 1994) to approximately 60 kDa in some monocots, such as barley and sugarcane (Vogel et al. 1999; Barthet and Hilu 2007). Examination of transcript and protein levels during development and after etiolation revealed a physiological role for MATK in the plant (Barthet and Hilu 2007). Although these findings indicate that this gene is expressed and functions in the plant, they do not demonstrate MATK activity as a group II intron maturase.

Indirect molecular evidence corroborates the putative maturase activity of MATK. Recombinant *matK* protein was shown to bind to the two group II intron containing transcripts, *trnK* and *trnG*, in gel-retardation assays, but not transcripts of *rps16*, which also contains a group II intron (Liere and Link 1995). MATK, therefore, may have multiple substrates for its activity but retains some specificity. Studies of the white barley mutant *albostrians* also support the suggestion that MATK functions as a chloroplast maturase with potentially multiple substrates for its activity. The *albostrians* mutant is deficient in chloroplast ribosome activity and, hence, lacks all chloroplast proteins, including MATK (Hagemann and Scholz 1962; Vogel et al. 1999). Six group II intron containing transcripts (*atpF*, *trnA*, *trnI*, *rpl2*, and *rps12 cis*) were reported to lack proper splicing in the *albostrians* mutant (Hess et al. 1994; Hübschmann et al. 1996; Vogel et al. 1997, 1999). In conflict with the previous work of Liere and Link (1995), the precursor transcript for *trnG* was efficiently processed in the *albostrians* mutant (Vogel et al. 1999), indicating that the maturase for this transcript is nuclear-encoded. Two transcripts identified to be unspliced in the *albostrians* mutant (*rpl2*, and *rps12 cis*), however, are in agreement with those predicted by evolutionary studies on *E. virginiana* to require MATK.

Nearly all transcripts determined to require a chloroplast factor (MATK?) for intron processing in the *albostrians* mutant are involved in protein translation, a point that implies a possible critical role for MATK in chloroplast function. The exception is *atpF*. The protein product from this gene is not involved in translation but in the synthesis of ATP by oxidative phosphorylation of the proton gradient formed by the light-dependent reactions of photosynthesis. The expression of *matK* at both the RNA and the protein level is increased by light (Nakamura et al. 2003; Barthet and Hilu 2007). The influence of light on *matK* expression may have evolved for regulation of proteins involved in the light-dependent reactions of photosynthesis such as *atpF* by preventing or accelerating, respectively, intron processing in these transcripts.

10.7 Conclusions

The relatively high rate of substitution in the plastid *matK* gene both at the nucleotide and at the amino acid level and the notable high frequency of indels point to the unusual mode and tempo of evolution of this gene. The high rate of substitution in *matK* and its mode of evolution that approach neutrality can be construed as the underlying factors for the high phylogenetic signal in this gene compared with other genes used in plant phylogenetics. The expression of *matK* at both the RNA and the protein level in species from across land plants coupled with the maintenance of its ORF in plants that have lost *trnK* owing either to inversions or to elimination of a large proportion of plastid genes support the importance of *matK* in the chloroplast. The presence of structural features in *matK* secondary structure which are homologous to group II intron maturases support the proposed function of this gene as a group II intron maturase. The *matK* gene remains as an invaluable gene in plant phylogenetic reconstruction and as a model for studying gene evolution.

Acknowledgements We thank Sunny Crawley for comments on the manuscript. Work was supported by grants from the National Science Foundation (EF-043105) to K.W.H, and the Botanical Society of America, Sigma Xi, and the Virginia Academy of Science to M.M.B.

References

APG II (2003) An update of the Angiosperm Phylogeny Group classification for the orders and families of flowering plants: APG II. Bot J Linn Soc 141:399–436

Barthet MM, Hilu KW (2007) Expression of *matK*: functional and evolutionary implications. Am J Bot 94:1402–1412

Barthet MM, Hilu KW (2008) Evaluating evolutionary constraint on the rapidly evolving gene *matK* using protein composition. J Mol Evol. 66:85–97

Chase MW, Soltis DE, Olmstead RG, Morgan D, Les DH, Mishler BD, Duvall MR, Price R, Hills HG, Qui Y-L, Kron KA, Rettig JH, Conti E, Palmer JD, Manhart JR, Sytsma KJ, Michaels HJ, Kress WJ, Karol KG, Clark WD, Hedren M, Gaut BS, Jansen RK, Kim K-J, Wimpee CF, Smith JF, Furnier GR, Strauss SH, Xiang QY, Plunkett GM, Soltis PS, Williams SE, Gadek PA, Quinn CJ, Eguiarte LE, Golenberg E, Learn GH, Graham S, Barrett SCH, Dayanandan S, Albert VA (1993) Phylogenetics of seed plants: an analysis of nucleotide sequences from the plastid gene *rbcL*. Ann Miss Bot Gard 80:528–580

Cronquist A (1988) The evolution and classification of flowering plants, 2nd edn. New York Botanical Garden, New York

Cui X, Matsuura M, Wang Q, Ma H, Lambowitz AM (2004) A group II intron-encoded maturase functions preferentially *in cis* and requires both the reverse transcriptase and X domains to promote RNA splicing. J Mol Biol 340:211–231

Daniell H, Lee S-B, Grevich J, Saski C, Quesada-Vargas T, Guda C, Tomkins J, Jansen RK (2006) Complete chloroplast genome sequences of *Solanum bulbocastanum, Solanum lycopersicum* and comparative analyses with other Solanaceae genomes. Theor Appl Genet 112:1503–1518

du Jardin P, Portetelle D, Harvengt L, Dumont M, Wathelet B (1994) Expression of intron-encoded maturase-like polypeptides in potato chloroplasts. Curr Genet 25:158–163

Eickbush TH, Malik HS (2002) Origins and evolution of retrotransposons. In: Craig NL et al (eds) Mobile DNA II. American Society for Microbiology, Washington, pp 1111–1144

Ems SC, Morden CW, Dixon CK, Wolfe KH, dePamphilis CW, Palmer JD (1995) Transcription, splicing and editing of plastid RNAs in the nonphotosynthetic plant *Epifagus virginiana*. Plant Mol Biol 29:721–733

Farris JS (1977) Phylogenetic analysis under Dollo's law. Syst Zool 26:77–88

Graham SW, Olmstead RG (2000) Utility of 17 chloroplast genes for inferring the phylogeny of the basal angiosperms. Am J Bot 87:1712–1730

Hagemann R, Scholz F (1962) Ein Fall Gen-induzierter Mutationen des Plasmotyps bei Gerste. Zuchter 32:50–59

Hausner G, Olson R, Simon D, Johnson I, Sanders ER, Karol KG, McCourt RM, Zimmerly S (2006) Origin and evolution of the chloroplast *trnK* (*matK*) intron: a model for evolution of group II intron RNA structures. Mol Biol Evol 23:380–391

Hess WR, Hoch B, Zelts P, Hübschmann T, Kössel H, Börner T (1994) Inefficient *rpl2* splicing in barley mutants with ribosome-deficient plastids. Plant Cell 6:1455–1465

Hilu KW, Alice LA (1999) Evolutionary implications of *matK* indels in Poaceae. Am J Bot 86:1735–1741

Hilu KW, Liang H (1997) The *matK* gene: sequence variation and application in plant systematics. Am J Bot 84:830–839

Hilu KW, Alice LA Liang H (1999) Phylogeny of Poaceae inferred from *matK* sequences. Ann Mo Bot Gard 86:835–851

Hilu KW, Borsch T, Müller K, Soltis DE, Soltis PS, Savolainen V, Chase MW, Powell MP, Alice LA, Evans R, Sauquet H, Neinhuis C, Slotta TAB, Jens GR, Campbell CS, Chatrou LW (2003) Angiosperm phylogeny based on *matK* sequence information. Am J Bot 90:1758–1776

Hübschmann T, Hess WR, Börner T (1996) Impaired splicing of the *rps12* transcript in ribosome-deficient plastids. Plant Mol Biol 30:109–123

Johnson LA, Soltis DE (1994) MatK DNA sequences and phylogenetic reconstruction in Saxifragaceae s. str. Syst Bot 19:143–156

Källersjö M, Albert VA, Farris JS (1999) Homoplasy increases phylogenetic structure. Cladistics 15:91–93

Li WH, Graur D (1991) Fundamentals of molecular evolution. Sinauer, Sunderland

Liang H, Hilu KW (1996) Application of the *matK* gene sequences to grass systematics. Can J Bot 74:125–134

Liere K, Link G (1995) RNA-binding activity of the *matK* protein encoded by the chloroplast *trnK* intron from mustard (*Sinapis alba* L.). Nucleic Acids Res 23:917–921

Mathews S, Donoghue MJ (2000) Basal angiosperm phylogeny inferred from duplicate phytochromes A and C. Int J Plant Sci 161(Suppl):41–55

Mohr G, Perlman PS, Lambowitz AM (1993) Evolutionary relationships among group II intron-encoded proteins and identification of a conserved domain that may be related to maturase function. Nucleic Acids Res 21:4991–4997

Moran JV, Mecklenburg, Sass P, Belcher SM, Mahnke D, Lewin A, Perlman PS (1994) Splicing defective mutants of the COXI gene of yeast mitochondrial DNA: initial definition of the maturase domain of the group II intron aI2. Nucleic Acids Res 22:2057–2064

Müller KF, Borsch T, Hilu KW (2006) Phylogenetic utility of rapidly evolving DNA at high taxonomical levels: contrasting *matK*, *trnT-F* and *rbcL* in basal angiosperms. Mol Phylogenet Evol 41:99–117

Nakamura T, Furuhashi Y, Hasegawa K, Hashimoto H, Watanabe K (2003) Array-based analysis on tobacco plastid transcripts: preparation of a genomic microarray containing all genes and all intergenic regions. Plant Cell Phys 44:861–867

Nandi OI, Chase MW, Endress PL (1998) A combined cladistic analysis of angiosperms using *rbcL* and non-molecular data sets. Ann Mo Bot Gard 85:137–212

Neuhaus H, Link G (1987) The chloroplast tRNA^{Lys} (UUU) gene from mustard (*Sinapsis alba*) contains a class II intron potentially coding for a maturase-related polypeptide. Curr Genet 11:251–257

Noah JW, Lambowitz AM (2003) Effects of maturase binding and Mg^{2+} concentration on group II intron RNA folding investigated by UV cross-linking. Biochemistry 42:12466–12480

Olmstead RG, Palmer JD (1994) Chloroplast DNA systematics: a review of methods and data analysis. Am J Bot 81:1205–1224

Ophir R, Itoh T, Graur D, Gojobori T (1999) A simple method for estimating the intensity of purifying selection in protein-coding genes. Mol Biol Evol 16:49–53

Pryer KM, Schneider H, Smith AR, Cranfill R, Wolf PG, Hunt JS, Sipes SD (2001) Horsetails and ferns are a monophyletic group and the closest living relatives to seed plants. Nature 409:618–622

Qiu Y-L et al (2006) The deepest divergences in land plants inferred from phylogenmic evidence. Proc Natl Acad Sci USA 103:15511–15516

Quint M, Classen-Bockhoff R (2006) Phylogeny of Bruniaceae based on *matK* and ITS sequence data. Int J Plant Sci 167:135–146

Rambo RP, Doudna JA (2004) Assembly of an active group II intron-maturase complex by protein dimerization. Biochemistry 43:6486–6497

Saldanha R, Mohr G, Belfort M, Lambowitz AM (1993) Group I and group II introns. FASEB J 7:15–24

Saldanha R, Chen B, Wank H, Matsuura M, Edwards J, Lambowitz AM (1999) RNA and protein catalysis in group II intron splicing and mobility reactions using purified components. Biochemistry 38:9069–9083

Savolainen V, Chase MW, Hoot SB, Morton CM, Soltis DE, Bayer C, Fay MF, De Bruijn AY, Sullivan S, Qiu Y-L (2000) Phylogenetics of flowering plants based upon a combined analysis of plastid *atpB* and *rbcL* gene sequences. Syst Biol 49:306–362

Soltis DE, Soltis PS (1998) Choosing an approach and an appropriate gene for phylogenetic analysis. In: Soltis PS, Soltis DE, Doyle JJ (eds) Molecular systematics of plants II: DNA sequencing. Kluwer, Boston, pp 2–31

Soltis PS, Soltis DE, Chase MW, Mort ME, Albach DC, Zanis M, Savolainen V, Hahn W, Hoot SB, Fay MF, Axtell M, Swensen SM, Prince LM, Kress WL, Nixon KC, JS Farris (2000)

Angiosperm phylogeny inferred from 18S rDNA, *rbcL*, and *atpB* sequences. Bot J Linn Soc 133:381–461

Steane DA (2005) Complete nucleotide sequence of the chloroplast genome from the Tasmanian blue gum, *Eucalyptus globulus* (Myrtaceae). DNA Res 12:215–220

Sugita M, Shinozaki K, Sugiura M (1985) Tobacco chloroplast tRNALys (UUU) gene contains a 2.5-kilobase-pair intron: an open reading frame and a conserved boundary sequence in the intron. Proc Natl Acad Sci USA 82:3557–3561

Swofford, DL, Olsen GJ, Waddell PJ, Hillis D (1996) Phylogenetic inference. In: Hillis DM, Moritz C, Mable BK (eds) Molecular systematics, 2nd edn. Sinauer, Sunderland, pp 407–514

Vogel J, Hubschmann T, Borner T, Hess WR (1997) Splicing and intron-internal RNA editing of *trnK-matK* transcripts in barley plastids: support for MatK as an essential splicing factor. J Mol Biol 270:179–187

Vogel J, Borner T, Hess W (1999) Comparative analysis of splicing of the complete set of chloroplast group II introns in three higher plant mutants. Nucleic Acids Res 27:3866–3874

Whitten WM, Williams NH, Chase MW (2000) Subtribal and generic relationships of Maxillarieae (Orchidaceae) with emphasis on Stanhopeinae: combined molecular evidence. Am J Bot 87:1842–1856

Wolf PG, Rowe CA, Sinclair RB, Hasebe M (2003) Complete nucleotide sequence of the chloroplast genome from a leptosporangiate fern, *Adiantum capillus-veneris* L. DNA Res 10:59–65

Wolf PG, Rowe CA, Hasebe M (2004) High levels of RNA editing in a vascular plant chloroplast genome: analysis of transcripts from the fern *Adiantum capillus-veneris*. Gene 339:89–97

Wolfe, KH, Morden CW, Palmer JD (1992) Function and evolution of a minimal plastid genome from a nonphotosynthetic parasitic plant. Proc Natl Acad Sciences, USA 89:10648–10652

Xiang Q-Y, Soltis DE, Morgan DR, Soltis PS (1998) Phylogenetic relationships of Cornaceae and close relatives inferred from *matK* and *rbcL* sequences. Am J Bot 85:285–297

Young ND, dePamphilis CW (2000) Purifying selection detected in the plastid gene *matK* and flanking ribozyme regions within a group II intron of nonphotosynthetic plants Mol Biol Evol 17:1933–1941

Young ND, dePamphilis CW (2005) Rate variation in parasitic plants: correlated and uncorrelated patterns among plastid genes of different function. BMC Evol Biol 5:16–25

Zanis M, Soltis DE, Soltis PS, Mathews S, Donoghue MJ (2002) The root of angiosperms revisited. Proc Natl Acad Sci USA 99:6848–6853

Chapter 11
Phylogeography and Conservation of the Rare South African Fruit Chafer *Ichnestoma stobbiai* (Coleoptera: Scarabaeidae)

Ute Kryger and Clarke H. Scholtz

Abstract *Ichnestoma stobbiai* is an endangered fruit chafer (Scarabaeidae: Cetoniinae) that occurs in small habitat fragments of South Africa. The adults of this species are short-lived and the females are flightless. Thus, the vagility of these beetles is extremely low. Prompted by the recent discovery of morphological divergence between geographic populations, this genetic study aimed to assess genetic differentiation within and among these different populations. DNA sequencing of the cytochrome c oxidase subunit 1 mitochondrial gene was used to determine the genetic composition of the populations. Most populations revealed low haplotype diversity. Phylogenetic analysis of the sequence data resulted in a basal polytomy. Nested clade analysis inferred allopatric fragmentation for all significant clades. This reconfirms the original hypothesis that the extant populations represent relicts of a single, formerly widely distributed species. All habitat patches should be protected and a detailed plan for genetic augmentation should be worked out.

11.1 Introduction

Human activity directly and indirectly impacts on the biological diversity of the planet. As a consequence of human-induced habitat fragmentation many animal populations decrease in range and size and the concurrent loss of genetic diversity reduces their ability to adapt to stochastic environmental changes (Frankham et al. 2002).

While the welfare of many of the charismatic megafauna species receives a lot of public attention, the fate of smaller animals such as invertebrates is still mostly deemed insignificant, despite the fact that Edward O. Wilson, one of the most prominent figures in the field of biodiversity conservation, promotes insects as "the

---

U. Kryger and C.H. Scholtz
Department of Zoology and Entomology, University of Pretoria, Pretoria 0002, South Africa
ukryger@zoology.up.ac.za, chscholtz@zoology.up.ac.za

Fig. 11.1 Map of South Africa indicating the geographic position of Gauteng province

small things that run the world" (speech at his acceptance of the TED 2007 Prize). A pleasant exception to the trend of ignoring the importance of invertebrates in biodiversity management is the systematic conservation plan that is used by the local conservation agency Gauteng Department of Agriculture, Conservation and Environment (GDACE) in Gauteng province, South Africa (Fig. 11.1). This small province, which includes the metropolitan areas of Johannesburg and Pretoria, is one of the economic development hubs in South Africa. Effective biodiversity conservation is of particular relevance there, since landscapes are being fragmented and transformed at a rapid rate. In the conservation plan mentioned, several terrestrial arthropod species were identified to be of potential conservation concern. One beetle, *Ichnestoma stobbiai* Holm 1992, was recognized as a conservation priority. The genus *Ichnestoma* Gory & Percheron 1833 is considered one of the more primitive and atypical genera of the tribe Cetoniini (Scarabaeidae: Cetoniinae) and comprises 13 known species that are restricted to the older mountain ranges of South Africa (Holm and Marais 1992, Perissinotto et al. 1999). *I. stobbiai* in particular occurs only in small fragments of pristine grassland along the Transvaal Magaliesberg system. This rare fruit chafer is mostly endemic to Gauteng province, with a single population occurring in the adjacent parts of North West province. Currently approximately 30% of the 2,171 km^2 of the species' extent of occurrence is transformed (Ian Engelbrecht, GDACE, personal communication); four previously known local populations have already been destroyed. It must be expected that the pressures of high human population densities in the relevant areas will lead to further transformation, degradation and fragmentation of habitat in the near future. This will put *I. stobbiai* under severe pressure because with its highly specialized biology it is extremely vulnerable to habitat changes (Davis et al. 2005).

The adults of these beetles have no functional mouthparts and do not feed. The first soaking spring rains (more than 12 mm) enable the adult *I. stobbiai* to break through their clay cocoons and the males appear simultaneously in mating swarm flights for a few days if the weather is sunny. The sole activity of their adult lives lies in reproduction. The females are flightless and shelter under grass tufts while releasing pheromones that attract the flying males. Hence, the dispersal capabilities of both sexes are very limited. Because of their limited energy reserves (missing mouthparts), the males cannot fly over long distances and the flightless females are extremely philopatric, sometimes not even fully emerging before being located by males and mating underground. Sometimes several males locate the same female and all follow her underground. Shortly after mating and depositing the eggs, they die. The three larval instars live subterraneanly, which is thought to be an adaptation to the frequent fires in the grassland habitat. They remain in the immediate vicinity of the oviposition site and probably feed on detritus for 9 months and then pupate 10–15 cm below the soil surface.

The extinction risk is generally believed to be particularly high in taxa with small populations that occur over small geographic ranges and that have a slow life history (Purvis et al. 2000). All this holds true for *I. stobbiai* and it is considered to be one of the most threatened animal species in the province. This has led to it currently being assessed for inclusion in the IUCN Red List (IUCN Standards and Petitions Working Group 2006) as "Endangered B1 + 2ab" (Ian Engelbrecht, GDACE, personal communication).

The original description of *I. stobbiai* was provided by Holm (1992), but subsequently it has been noted that there is substantial morphological variation between separated populations of the species (James Harrison, personal communication). This raised the question of whether the fragmentation into isolated small habitat pockets in this species is indeed recent and human-induced, or whether it represents a natural and historical pattern that had already led to allopatric speciation within the complex.

In the process of expert consultation, GDACE had approached our working group to investigate the genetic structure of this species, in order to determine whether all geographically isolated populations should be managed as one conservation unit. Here we present the results for the genetic variability within and between the few known localities of *I. stobbiai* and elucidate the phylogeographic history of the species.

11.2 Materials and Methods

11.2.1 Data Production

Over the duration of two consecutive years we managed to sample specimens from nine localities (Fig. 11.2). We omit the precise GPS data here on purpose, because

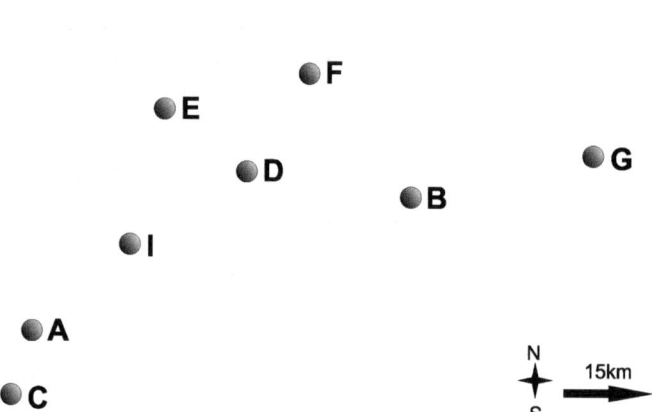

Fig. 11.2 The geographic orientation of the nine *Ichnestoma stobbiai* localities and their geographic distance from each other

Table 11.1 Sampled specimens per locality (n), the number of haplotypes, haplotype diversity (h) and nucleotide diversity (π)

Localities	n	No. of haplotypes	h	π
A	4	4	1	0.0102
B	9	2	0.222	0.0007
C	1	1	NA	NA
D	12	2	0.303	0.0005
E	12	3	0.439	0.0008
F	3	3	1	0.0055
G	4	1	0	0
H	9	1	0	0
I	3	2	0.667	0.0066
NA				

I. stobbiai is also very popular among beetle collectors; in order to prevent further damage to the already small population sizes all locality data must be kept confidential. The numbers of specimens sampled and analyzed per locality are shown in Table 11.1. As an outgroup we included one specimen of another South African fruit chafer species, *Raceloma jansoni*. The beetles were collected in the field and immediately transferred into 99% ethanol; after 24–48 h the ethanol was replaced. The samples were stored at −20°C.

DNA extractions were done from thoracic muscle tissue or legs using the High Pure polymerase chain reaction (PCR) template preparation kit (Roche, Mannheim Germany).

The cytochrome *c* oxidase subunit 1 (COI) gene was selected as the mitochondrial DNA marker in this study. Past work has indicated that sequence divergences at

COI regularly enable the discrimination of closely allied species in all animal phyla except Cnidaria (Hebert et al. 2003). The COI marker has shown substantial variation within the Coleopteran family Scarabaeidae at inter- and intraspecific levels (Roslin 2001; Clarke et al. 2001, Cabrero-Sanudo and Zardoya 2003; Miller and Allsopp 2005, Sole et al. 2005).

The primer sequences used for DNA amplification of a 1,345 base pair (bp) long fragment of COI were obtained from Simon et al. (1994). These were TL2-N-3014 and C1-J-1718. PCR was performed in a final volume of 50 μl using a PerkinElmer Gene Amp 2400. This reaction mixture contained 5 μl of 10 × buffer, 4 μl of 2.5 mM dNTPs, 20 pmol of each primer, 36 μl of double-distilled H_2O, 1 unit of *Taq* DNA polymerase (Biotools, Madrid, Spain) and 5 μl of DNA template. Thermal cycling parameters comprised an initial denaturation step for 90 s at 94°C followed by 30 cycles at 94°C for 22 s, 48°C for 30 s and 72°C for 90 s with a final elongation step at 72°C for 1 min (Sole et al. 2005). PCR products were visualized on 1.5% agarose gels run in tris(hydroxymethyl)aminomethane–acetate–EDTA buffer and stained with ethidium bromide prior to DNA purification.

Amplified COI gene products were purified using the High Pure PCR product purification kit (Roche, Mannheim, Germany) following the manufacturer's specifications. An aliquot of the reaction mixture was subjected to agarose gel electrophoresis for verification of PCR amplification prior to sequencing. Sequencing reactions were performed with version 3.1 of the ABI Prism Big Dye terminator cycle sequencing ready reaction kit (PerkinElmer, Foster City, USA). The following sequencing profile was used: 96°C for 5 min; 25 cycles of 96°C for 10 s, 50°C for 5 s and 60°C for 4 min; with holding at 4°C. Purified PCR fragments were sequenced using the forward primer C1-J-1718 and the reverse primer C1-N-2757 (Monteiro and Pierce 2000). Sequencing was carried out in 10-μl final volume reactions, consisting of either of these primers at 3.2 pmol, 2 μl BigDye^{TM} terminator reaction mix, 1 μl 5 × buffer, x μl DNA template and $(10 - x)$ μl Sabax water. The amount of DNA template (x) used varied depending on its concentration after purification. This was determined by visualizing a 2-μl aliquot of DNA template and estimating its concentration in relation to 2 μl of 12.5 μg/μl Lambda. BigDye cycle sequencing conditions provided by the GeneAmp^{TM} PCR system 9700 (Applied Biosystems, Foster City, CA, USA) were used. Sequences were generated using an ABI 3100 automated sequencer (PerkinElmer, Foster City, USA).

11.2.2 Data Analyses

The corresponding L- and H-primer sequences of each individual were compared and edited in Chromas Lite (version 2.0, Technelysium, 1998–2004). The consensus sequences of all individuals were then automatically aligned with the program Clustal X (Thompson et al. 1997) and these alignments were manually optimized. Analyses to retrieve phylogenetic associations among lineages were performed in Paup*4.0 beta 10 version (Swofford 2002). We ran a maximum likelihood ratio test as implemented in Modeltest version 3.04 (Posada and Crandall 1998) in order to determine the model of nucleotide substitution that fits the presented data most adequately.

A gene tree containing all the haplotypes was constructed with the neighbor-joining algorithm (Saitou and Nei 1987) based on pairwise distances and using the option of randomized tie breaking. Nodal support for the tree was assessed by 1,000 bootstrap replicates (Felsenstein 1985; Hillis and Bull 1993).

We performed 100 heuristic searches for the parsimony analysis (Felsenstein 1982, 1988), obtaining the starting trees by random stepwise addition (ten replicates) and swapping the branches via the tree-bisection–reconnection function. Branch support was assessed with 1,000 bootstrap replicates.

The program DnaSP version 3.53 (Rozas and Rozas 1999) was employed to calculate haplotype diversity (H; Nei and Tajima 1981), nucleotide diversity (π; Nei 1987), a neutrality test in the form of Tajima's D (Tajima 1989), pairwise F_{ST} values and Nm values (after Hudson et al. 1992).

We estimated a 95% parsimony cladogram for the mitochondrial DNA (mtDNA) CR-I data in TCS (Clement et al. 2000) treating gaps as a fifth character state. Loops (i.e., reticulations) in the network, which resulted from the homoplasy in the data, were broken in accordance with the predictions derived from coalescent theory by Donelly and Tavare (1986), Crandall and Templeton (1993) and Posada and Crandall (2001): (1) common haplotypes are more likely to be found at interior nodes of a cladogram, and rare haplotypes at the tips; (2) haplotypes represented by only one individual (singletons) are more likely connected to haplotypes from the same population than to haplotypes from different populations.

On the basis of this cladogram, the nested clade structure for our data set was determined following the nesting rules described in Templeton et al. (1987) and Templeton and Sing (1993).

A nested clade analysis (NCA) applying the methods described in Templeton et al. (1995) was performed using the program GeoDis (Posada et al. 2000).

The results of the NCA were interpreted according to Templeton et al. (1995), Crandall and Templeton (1996), and the new NCA key inference (2003), which was downloaded from the NCA Web page (http://zoology.byu.edu/crandall_lab/geodis.htm).

11.3 Results

The results presented are preliminary, the effort to sample more specimens from new localities is ongoing and the target is to collect and sequence 20 specimens per known locality.

During PCR a 1,345-bp-long fragment of the mitochondrial COI was amplified. The amplifications generally worked well and resulted in one single, crisp band. Sequencing with the forward and internal reverse primer resulted in a 700-bp-long fragment of COI.

The electropherograms were clear and the sequences of the forward and reverse primer corresponded perfectly to each other. A translation of the nucleotide sequences obtained into amino acid sequences using the software program MacClade (Maddison and Maddison 2002) did not contain any stop codons and resulted in the correct polypeptide. This was interpreted as evidence that the sequences produced were indeed of mitochondrial COI origin and not nuclear pseudogenes (Zhang and Hewitt 1996). A negative value for Tajima's D (-0.74760) was nonsignificant ($P > 0.10$) and therefore the sequences were taken to be selectively neutral and hence useful for phylogenetic and phylogeographic inferences.

Among the 56 in-group specimens sequenced, 20 different haplotypes were found. Table 11.1 shows the number of haplotypes per locality, Table 11.2 shows which specimens had the same haplotypes. There was one shared haplotype between localities B and C; and at locality C only one specimen has been found and sequenced so far. As becomes obvious from the rather low haplotype diversities and nucleotide diversities, five of the localities (B, D, E, G, H) are characterized by really low levels of genetic variability (Table 11.1). The pairwise F_{ST} values were generally very high (around 0.9, data not shown) and the Nm values were very low (close to zero migrants per generation, data not shown). Notable exceptions to this trend were the lower pairwise F_{ST} values between localities B and I (0.13) and between localities E and F (0.55) and the relatively high Nm value of 1.78 migrants per generation between localities B and I.

In Modeltest the HKY85 model (Hasegawa et al. 1985) with gamma correction and proportion of invariable sites (Gu and Zhang 1997) obtained the best

Table 11.2 Specimens and their haplotypes

Haplotype	Specimens with this haplotype
1	A3
2	A4
3	A6
4	A1
5	B4, B9, B10, B13, B14, B15, B16, C1
6	B8
7	I2, I3
8	I1
9	D1, D4, D5, D6, D8, D9, D10, D11, D12
10	D3, D7
11	D2
12	E3, E6
13	E1, E2, E4, E7, E8, E10, E11, E12
14	E9
15	E5
16	F4
17	F5
18	F11
19	G1, G5, G6, G7
20	H2, H3, H4, H5, H6, H8, H9, H10, H12

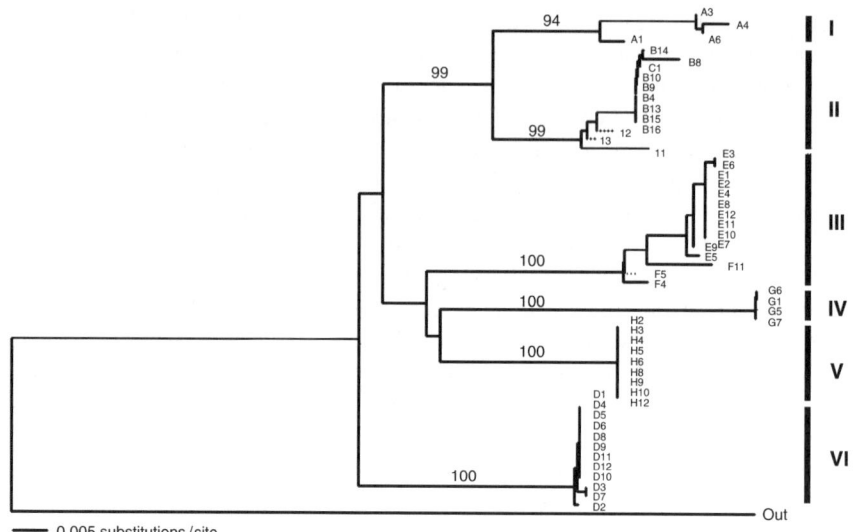

Fig. 11.3 Topology retrieved via a neighbor-joining analysis of 700 bas pairs of mitochondrial DNA (mtDNA) cytochrome *c* oxidase subunit 1 (COI) in Paup. Values *above the branches* represent the bootstrap support values after 1,000 replications. The *letters at the tips* represent the different specimens, with the coding corresponding to the locality labels in Fig. 11.2. The six genetic lineages are indicated via the *black bars with Roman numerals on the right*

likelihood score with a statistical significance of $P = 0.01$ and was thus selected for the neighbor-joining analysis. The tree topology resulting from the neighbor-joining analysis is shown in Fig. 11.3. Within the in-group, six different lineages were retrieved with very high bootstrap support values (Fig. 11.3). The neighbor-joining tree showed very short branches for samples of five localities, an indication of a pronounced lack of genetic variability within these localities. A notable exception was lineage I, which corresponds to a large population (many hundreds of individuals were observed on the collection day) sampled from a nature reserve. Assemblages II and III consisted of specimens from three and two different localities, respectively (compare Figs. 11.2, 11.3 for details).

The uncorrected sequence divergences (p) are shown in Table 11.3. Generally, within-lineage divergences ranged from 0 to 2.22% and between-lineage divergences ranged from 5.3 to 8.5%.

The parsimony analysis on the 20 different mtDNA COI haplotypes resulted in 108 informative characters and 14 best trees with 219 steps (CI = 0.68). Figure 11.4 shows the topology of the strict consensus tree with a basal polytomy and all bootstrap support values over 50%. This procedure recovered the same six lineages as in the neighbor-joining analysis (Fig. 11.3).

The unrooted haplotype network constructed in TCS is shown in Fig. 11.5. The six separate subnetworks corresponded to the six assemblages found in the neighbor-joining and parsimony analyses. The 95% parsimony probability threshold for linking the haplotypes with each other was 11 steps. The six subnetworks were

Table 11.3 Intra- and interlineage uncorrected (*p*) sequence divergences based on 700 base pairs of mitochondrial cytochrome *c* oxidase subunit 1 sequence

Lineage	I	II	III	IV	V	VI
I	2.00					
II	5.29	2.22				
III	7.96	7.88	1.71			
IV	7.66	8.46	7.14	0.00		
V	6.39	6.13	6.00	6.14	0.00	
VI	7.00	6.87	7.14	7.71	6.00	0.29

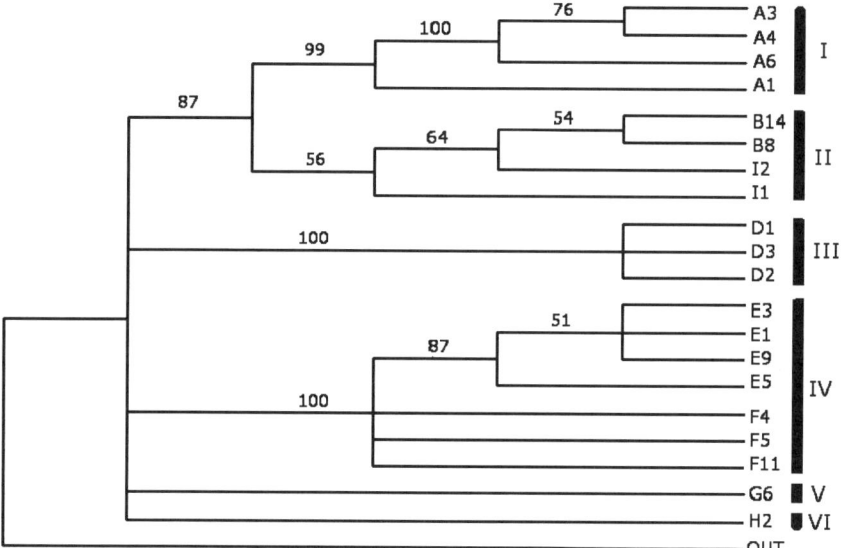

Fig. 11.4 Topology retrieved via a parsimony analysis of 700 base pairs of mtDNA COI in Paup. The *letters at the tips* signify the specimens representing the 20 haplotypes (compare with Table 11.2), with the coding corresponding to the locality labels in Fig. 11.2. The six genetic lineages are indicated via the *black bars with Roman numerals on the right*

separated from each other by 39–58 mutational steps and could thus not be linked within the parsimony parameters, a result that was also reflected in the basal polytomy in the parsimony analysis in Paup. Figure 11.5 also illustrates the nested clade structure imposed upon the unrooted network following the rules of Templeton et al. (1987) and Templeton and Sing (1993). The only four-step clade (4-1) contained two three-step clades that were actually separated by more than one step, but in order to simplify the nesting design the missing step levels that consisted of missing haplotypes only were collapsed.

Table 11.4 shows the phylogeographic inference chains for the three clades with statistically significant geographical structure. The geographical arrangement of haplotypes in clade 3-3 (containing specimens from localities E and F) cannot be conclusively interpreted following the 2003 inference key; however, based on the

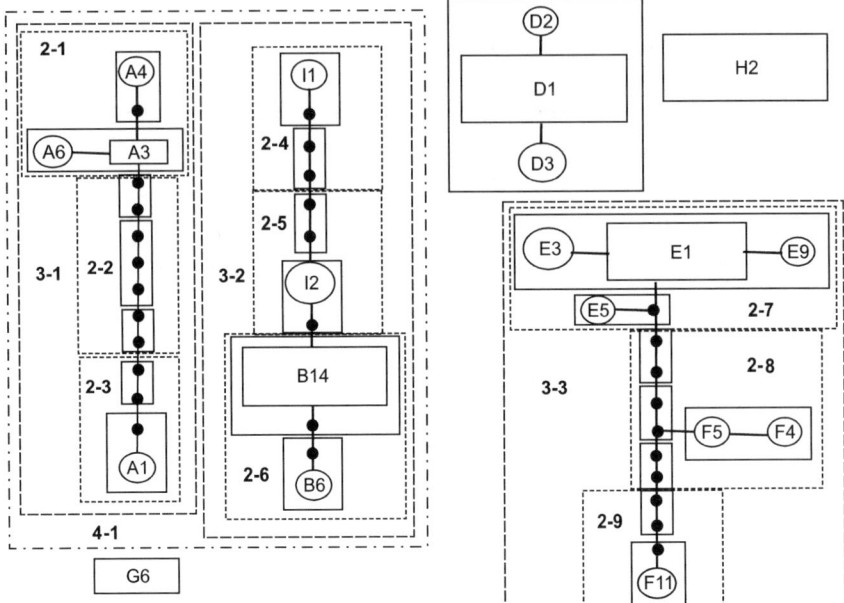

Fig. 11.5 Unrooted haplotype network based on 700 base pairs of mtDNA COI produced in TCS with the associated nested clade structure manually designed, following the rules of Templeton et al. (1987) and Templeton and Sing (1993). Haplotypes are represented by *circles* or *squares* (for haplotype numbers refer to Table 11.1), with the *squares* representing the ancestral haplotypes; the size of the symbols is proportionate to the frequencies of the haplotypes. Intermediate, missing haplotypes are illustrated as *black circles*, each connection represents one mutational step. One-step clades are represented by *black boxes with unbroken lines*, two-step clades by *boxes with a dotted line*, and three-step clades by *boxes with dashed lines*. The single four-step clade is represented by a *dashed–dotted line* and the actual number of steps between the two nested clades was 28 mutational steps. All two-step, three-step and four-step clades are numbered

Table 11.4 Clades with significant geographical structure ($P < 0.05$) and their phylogeographical interpretation following Templeton et al. (1995). See Fig. 11.5 for the composition of the clades and Fig. 11.2 for the localities. The numbers in the chain inference refer to the consecutive steps in the inference key with the final interpretation (see GeoDis Web page http://darwin.uvigo.es/download/geodisKey_11Nov05.pdf)

Clade	P	Chain inference
3-3	0.0000	1-2-11-17-NO inconclusive/isolation by distance
4-1	0.0020	1-19-NO allopatric fragmentation
Total cladogram	0.0000	1-19-NO allopatric fragmentation

2001 inference key it would be interpreted as the consequence of isolation by distance. According to the inference chain the geographical association of haplotypes in clade 4-1 and the entire cladogram was interpreted as the result of allopatric fragmentation, in both cases reinforced by the fact that the nested clades were separated from each other by a larger than average number of mutational steps.

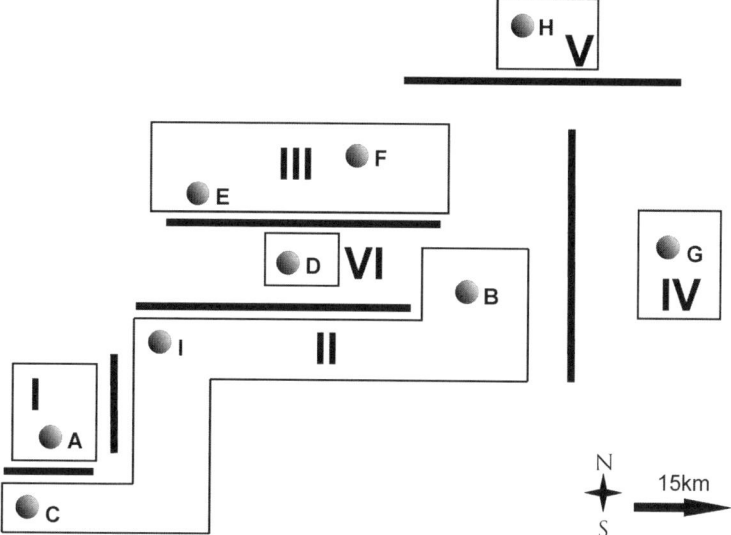

Fig. 11.6 The phylogeographic inference of "allopatric fragmentation" for *I. stobbiai* retrieved in GeoDis overlaid on the geographic map. The *black bars* signify fragmentation events. The *boxes with the Roman numerals* refer to the six genetic assemblages as defined by the Paup analyses (compare Figs. 11.3, 11.4)

Figure 11.6 illustrates the phylogeographic inferences obtained through GeoDis in overlay to the geography of the localities included, showing how allopatric fragmentation was the major factor shaping the current genetic structure found in the species.

11.4 Discussion

All analyses failed to resolve the internal, basal nodes of relationships among the localities of *I. stobbiai* investigated. Furthermore, all three analyses retrieved the same six genetic lineages within the total sample (Figs. 11.4–11.6). This gives this result a reasonable level of confidence.

The average pairwise COI sequence divergence found between the different lineages (6–8%) is comparable to the levels of divergence within other South African scarab species (unpublished results). Bell et al. (2003) found similarly high levels of intraspecific COI sequence divergences (up to 13.2% with a mean of 6.7%) and higher levels of COI sequence divergences between sister species of Australian dung beetles of the genus *Temnoplectron*, namely, 6.4–21.5%.

It is likely to be the genetic signature of a historically large population size and hence some retention of ancient polymorphism. In accordance with the reportedly rapid mutation rates for COI in Coleoptera (6% pairwise sequence divergence per million years; Juan et al. 1996, Clarke et al. 2001), our values of 6–8% would also be expected levels of intraspecific divergence. According to Hebert et al. (2003), the average interspecific COI sequence divergence between congeneric animal species is 11.3% and in an investigation of 891 beetle species pairs 85% of the comparisons showed interspecific COI sequence divergences of 8–16%. In relation to these literature values of related animal groups and together with the missing resolution in the phylogenetic analysis (Fig. 11.4), the levels of sequence divergence found within *I. stobbiai* do not support the idea of several species within the sample.

In the geographical association analysis, three clades showed significant association of genetic variation with geographic features and the phylogeographic inference of allopatric fragmentation corresponded well to the basal polytomy in the parsimony tree.

The occurrence of a shared mitochondrial COI haplotype in two individuals from assemblage II stemming from two different localities more than 60 km apart from each other represents the genetic signature of a previously panmictic population with a continuous distribution area spanning the entire area of assemblage II. It is impossible for adult *I. stobbiai* females to crawl over a distance of more than 60 km dispersing from one habitat island to the other (between locality B and locality C); therefore, the shared haplotype represents the signature of connectivity and gene flow in the past—and it places the currently observed fragmentation in the recent past. The basal polytomy in the parsimony tree from Paup supports the inference of the recent fragmentation of a previously continuous panmictic population (Neigel 2002). In the case of a recent fragmentation event not enough time has passed for lineage sorting to take place; if population sizes are small (as is the case for *I. stobbiai*), a basal polytomy hints that a very short time has passed since the actual fragmentation took place.

More evidence in support of this more recent dating of the fragmentation event came from the negative value for Tajima's D: had there been any long-term geographic subdivision between the populations of *I. stobbiai*, mutational differences would have accumulated among the six assemblages and Tajima's D would have a positive value. A negative Tajima's D, however, is consistent with a demographic history of expansion rather than historical subdivision (Marjoram and Donnelly 1994; Wakeley and Hey 1997; Knowles et al. 1999). Furthermore supporting this interpretation is the F_{ST}-based Nm value of 1.78 migrants per generation between localities B and I, which is again impossibly the signal of current female-mediated gene flow across the 45 km between the two localities in question, because the apterous females of *I. stobbiai* would hardly be able to crawl this distance within the short period of their adult lives. It is more likely the signal of a recent disruption of a previously connected and panmictic population over a continuous distribution area.

The otherwise high F_{ST} and correspondingly low Nm values would thus most likely be the consequence of most localities showing severely reduced genetic variability as a result of recent fragmentation into small, isolated habitat islands and resultant severe genetic drift. If these fragments were old and historically grown, they would have reached decent levels of genetic variability owing to a long time period in which they would have accumulated mutations. Furthermore, their placement in the parsimony topology (Fig. 11.4) would then be characterized by reciprocal monophyly. This is not the case. Hence, they do not fulfill the criteria for being treated as separate evolutionarily significant units (ESUs) *sensu* Moritz (1994, 1995). However, a nuclear marker should also be used to confirm this conclusion. Seeing that reciprocal monophyly for a mitochondrial marker is a necessary condition under the ESU definition of Moritz (1994), it can already be suggested, though, that the six assemblages within *I. stobbiai* should be managed as management units (Moritz 1994, 1995). Given the alarmingly low levels of genetic variability in five of the lineages (and their resulting vulnerability to environmental stochasticity), the possibility of translocations should be worked on. The logistics of doing so may be very complicated in a species that is only active for a couple of days each year; but they may be the only chance that these genetically depauperate populations have of escaping the extinction vortex (Frankham et al. 2002).

The utilization of a single mitochondrial marker, as well as a small sample size, in this study implies that no solid conservation strategies can be implemented at this point in time. Before such a decision can be made, it would be advisable to increase the sample size of the present study, carry out a similar analysis on a nuclear marker, and lastly increase efforts to discover any so far undiscovered populations. In ongoing research we are striving to get genetic data for 20 specimens per known locality. Furthermore, another mitochondrial and a nuclear marker are going to be employed in order to get a more complete picture of the present genetic variability in the species. This will also allow us to test for gene flow with the program Migrate (Beerli 1997–2001) and to calculate the time to the most recent common ancestor.

If, however, a conservation decision is to be made on the basis of this study only, it would be advisable to translocate individuals between populations so as to increase the evidently low level of genetic diversity within populations. As Burkey (1989, 1997) and Gonzalez et al. (1998) have shown, the fragmentation of once-continuous habitat blocks into multiple smaller populations increases the probability of extinction. This effect can be alleviated to some extent if there is dispersal among subpopulations in the fragmented landscape (Reed 2004). Given the extremely low dispersal capabilities of the fruit chafer of this study, translocation seems to be a valuable conservation management option.

A conservation decision also needs to take into consideration the fact that *I. stobbiai* persists in small habitat patches subject to succession and natural disturbance from which populations may disappear (Diogo et al. 1999). For this reason, a minimum number of populations are required for the long-term survival of the species. *I. stobbiai* is a species which is largely endemic to the Gauteng province of South Africa and it may indicate the existence of other endemic arthropods in this region. *I. stobbiai* may therefore be a useful focal taxon for conservation (Diogo et al. 1999), indicative of a relict assemblage of arthropods only able to persist within or near to Gauteng province. Protecting the remaining populations of this fruit chafer is therefore of high conservation priority and may be useful in protecting a larger assemblage of genetically unique components (Diogo et al. 1999).

11.5 Conclusion

The presence of a shared haplotype between two localities (B and C) separated by more than 60 km and the signal of female migration between two other localities (B and I) separated by more than 45 km, together with the lack of reciprocal monophyly in the phylogenetic tree (Fig. 11.4) and a negative value for Tajima's D suggest that the fragmentation of *I. stobbiai* into isolated geographic populations is of recent origin. Each of these tiny rest populations carries a substantial part of the evolutionary history and future potential of the species and therefore all known remnant localities are in need of immediate and rigorous protection if the species as a whole is to survive. This recommendation to the conservation authorities gains urgency from the fact that in recent years four previously known localities have already been transformed and the local populations of *I. stobbiai* became extinct. Within the framework of a possible effective translocation scheme of specimens between the remnant localities in order to genetically augment the localities with signatures of severe genetic drift (B, D, E, G and H) and therefore inbreeding, all rest localities are of utmost importance. In this context, the population on the nature reserve (locality A) could serve as a source population given its high census population size and the high level of genetic variability present there.

The future of this highly evolved and specialized fruit chafer species, like that of so many others, is seriously jeopardized by human activities motivated by profit and personal financial gain. It remains to be seen whether the influence of the relevant governmental nature conservation agencies is strong enough to protect this unique group of animals against the pressures of reckless developers, some of which have histories of blatantly ignoring any environmental impact assessment guidelines and erecting so-called eco-estates after bulldozing the relevant areas over and thus destroying any natural biodiversity present.

Acknowledgements We thank Christian Deschodt, Marcel Medrano and Vincent van der Merwe for help in the field and in the laboratory. We are grateful to the Gauteng Department for Agriculture and Environment for initiating this and partly funding this project. Grant support was through the National Research Foundation of South Africa.

References

Bell KL, Yeates DK, Moritz C, Monteith GB (2003) Molecular phylogeny and biogeography of the dung beetle genus *Temnoplectron* Westwood (Scarabaeidae: Scarabaeinae) from Australia's wet tropics. Mol Phylogenet Evol 31:741–753

Beerli P (1997–2001) Migrate: documentation and program, part of Lamarc. Version 1.1. Revised 30 April 2001. http://evolution.genetics.washington.edu/lamarc.html

Burkey TV (1989) Extinction in nature reserves: the effect of fragmentation and the importance of dispersal between reserve fragments. Oikos 55:75–81

Burkey TV (1997) Metapopulation extinction in fragmented landscapes: using bacteria and protozoa communities as model ecosystems. Am Nat 150:568–591

Cabrero-Sanudo F, Zardoya R (2003) Phylogenetic relationships of Iberian Aphodiini (Coleoptera: Scarabaeidae) based on morphological and molecular data. Mol Phylogenet Evol 31: 1084–1100

Clarke TE, Levin DB, Kavanaugh DH, Reimchen TE (2001) Rapid evolution in the *Nebria gregaria* group (Coleoptera: Carabidae) and the paleogeography of the Queen Charlotte islands. Evolution 55:1408–1418

Clement M, Posada D, Crandall KA (2000) TCS: a computer program to estimate gene genealogies. Mol Ecol 9:1657–1659

Crandall KA, Templeton AR (1993) Empirical tests of some predictions from coalescent theory with applications to intraspecific phylogeny reconstruction. Genetics 134:959–969

Crandall KA, Templeton AR (1996) Applications of intraspecific phylogenetics. In: Harvey PH, Leigh Brown AJ, Maynard Smith J, Nee S (eds) New uses for new phylogenies. Oxford University Press, New York, pp 81–99

Davis ALV, Scholtz CH, Deschodt C (2005) A dung beetle survey of selected Gauteng nature reserves: implications for conservation of the provincial scarabaeine fauna. Afr Entomol 13: 1–16

Diogo AC, Vogler AP, Gimenez A, Gallego D, Galian J (1999) Conservation genetics of *Cicindela deserticoloides*, an endangered tiger beetle endemic to southeastern Spain. J Insect Conserv 3:117–123

Donnelly P, Tavare S (1986) The ages of alleles and a coalescent. Adv Appl Probab 18:1–19

Felsenstein J (1982) Numerical methods for inferring evolutionary trees. Q Rev Biol 57:379–404

Felsenstein J (1985) Confidence limits on phylogenies: an approach using the bootstrap. Evolution 39:783–791

Felsenstein J (1988) Phylogenies from molecular sequences: inference and reliability. Annu Rev Genet 22:521–565

Frankham R, Ballou JD, Briscoe DA (2002) Introduction to conservation genetics. Cambridge University Press, Cambridge

Gonzalez A, Lawton JH, Gilbert FS, Blackburn TM, Evans-Freke I (1998) Meta-population dynamics, abundance, and distribution in a microecosystem. Science 281:2045–2047

Gu X, Zhang J (1997) A simple method for estimating the parameters of substitution rate variation among sites. Mol Biol Evol 14:1106–1113

Hasegawa M, Kishino H, Yano T-A (1985) Dating of the human-ape splitting by a molecular clock of mitochondrioal DNA. J Mol Evol 22:160–174

Hebert PDN, Ratnasingham S, DeWaard JR (2003) Barcoding animal life: cytochrome c oxidase subunit 1 divergences among closely related species. Proc R Soc Lond Ser B 270:96–99

Hillis DM, Bull JJ (1993) An empirical test of bootstrapping as a method for assessing confidence in phylogenetic analysis. Syst Biol 42:182–192

Holm E (1992) Revision of the African Cetoninae V: genus *Ichnestoma* Gory & Percheron (including *Gariep* Péringuey) (Coleoptera: Scarabaeidae). Ann Transvaal Mus 35: 374–376

Holm E, Marais E (1992) Fruit chafers of southern Africa. Sigma, Pretoria

Hudson RR, Slatkin M, Maddison WP (1992) Estimation of levels of gene flow from DNA sequence data. Genetics 132:583–589

IUCN Standards and Petitions Working Group (2006) Guidelines for using the IUCN Red List categories and criteria. Version 6.2. Prepared by the standards and petitions working group of the IUCN SSC biodiversity assessments sub-committee in December 2006. http://app.iucn.org/webfiles/doc/SSC/RedListGuidelines.pdf

Juan C, Ibrahim KM, Oromi P, Hewitt GM (1996) Mitochondrial DNA sequence variation of *Pimelia* darkling beetles on the island of Tenerife (Canary Islands). Heredity 77:589–598

Knowles LL, Futuyama DJ, Eanes WF (1999) Insight into speciation from historical demography in the phytophagous beetle genus *Ophraella*. Evolution 53:1846–1856

Maddison DR, Maddison WP (2002) McClade 4: analysis of phylogeny and character evolution, version 4.05. Sinauer, Sunderland

Marjoram P, Donnelly P (1994) Pairwise comparisons of mitochondrial DNA sequences in subdivided populations and implications for early human evolution. Genetics 136:673–683

Miller J, Allsopp PG (2005) Phylogeography of the scarab beetle *Antritrogus parvulus* Britton (Coleoptera: Scarabaeidae) in south-eastern Queensland, Australia. Aust J Entomol 38: 189–196

Monteiro A, Pierce NE (2000) Phylogeny of *Bicyclus* (Lepidoptera: Nymphalidae) inferred from COI, COII, and EF-1α gene sequences. Mol Phylogenet Evol 18:264–281

Moritz C (1994) Defining "evolutionarily significant units" for conservation. TREE 9:373–375

Moritz C (1995) Use of molecular phylogenies for conservation. Philos Trans R Soc Lond Ser B 349:113–118

Nei M (1987) Molecular evolutionary genetics. Columbia University Press, New York

Nei M, Tajima F (1981) DNA polymorphism detectable by restriction endonucleases. Genetics 97:145–163

Neigel JE (2002) Is F_{ST} obsolete? Conserv Genet 3:167–173

Perissinotto R, Smith TJ, Stobbia P (1999) Description of adult and larva of Ichnestoma pringlei n. sp. (Coleoptera Scarabaeidae Cetoniinae), with notes on its biology and ecology. Trop Zool 12:219–229

Posada D, Crandall KA (1998) Modeltest: testing the model of DNA substitution. Bioinformatics 14:817–818

Posada D, Crandall KA (2001) Intraspecific gene genealogies: trees grafting into networks. TREE 16:37–45

Posada D, Crandall KA, Templeton AR (2000) GeoDis: a program for the cladistic nested analysis of the geographical distribution of genetic haplotypes. Mol Ecol 9:487–488

Purvis A, Gittleman JL, Cowlishaw G, Mace GM (2000) Predicting extinction risk in declining species. Proc R Soc Lond Series B, Biologcial Sciences 267:1947–1952

Reed DH (2004) Extinction risk in fragmented habitats. Anim Conserv 7:181–191

Roslin T (2001) Spatial population structure in a patchily distributed beetle. Mol Ecol 10:823–837

Rozas J, Rozas R (1999) DnaSP version 3: an integrated program for molecular population genetics and molecular evolution analysis. Bioinformatics 15:174–175

Saitou N, Nei M (1987) The neighbor-joining method: a new method for reconstructing phylogenetic trees. Mol Biol Evol 4:406–425

Simon C, Frati F, Beckenbach AT, Crespi B, Liu H, Flook P (1994) Evolution, weighting, and phylogenetic utility of mitochondrial gene sequences and a compilation of conserved PCR primers. Ann Entomol Soc Am 87:651–701

Sole CL, Scholtz CH, Bastos AD (2005) Phylogeography of the Namib Desert dung beetles *Scarabaeus (Pachysoma)* MacLeay (Coleoptera: Scarabaeidae). J Biogeogr 32:75–84

Swofford DL (2002) Paup*. Phylogenetic analysis using parsimony (*and other methods). Version 4. Sinauer, Sunderland

Tajima F (1989) Statistical method for testing the neutral mutation hypothesis by DNA polymorphism. Genetics 129:585–595

Templeton AR, Sing CF (1993) A cladistic analysis of phenotypic associations with haplotypes inferred from restriction endonuclease mapping. IV. Nested analyses with cladogram uncertainty and recombination. Genetics 134:659–669

Templeton AR, Boerwinkle E, Sing CF (1987) A cladistic analysis of phenotypic associations with haplotypes inferred from restriction endonuclease mapping. I. Basic theory and an analysis of alcohol dehydrogenase activity in Drosophila. Genetics 117:343–351

Templeton AR, Routman E, Phillips CA (1995) Separating population structure from population history: a cladistic analysis of the geographical distribution of mitochondrial DNA haplotypes in the tiger salamander, *Ambystoma tigrinum*. Genetics 140:767–782

Thompson JD, Gibson TJ, Plewniak TF, Jeanmougin F, Higgins DG (1997) The CLUSTAL X windows interface: flexible strategies for multiple sequence alignment aided by quality analysis tools. Nucleic Acids Res 24:4876–4882

Wakeley J, Hey J (1997) Estimating ancestral population parameters. Genetics 145:847–855

Zhang DX, Hewitt GM (1996) Nuclear integrations: challenges for mitochondrial DNA markers. TREE 11:247–251

Chapter 12
Nothing in Medicine Makes Sense Except in the Light of Evolution: A Review

Bernard Swynghedauw

Abstract Applying evolutionary biology to medical problems is surprisingly new and still absent from medical as well biological teaching, despite the fact that "nothing in biology makes sense except in the light of evolution" (T. Dobzhansky in *Am. Biol. Teach.* 35:125, 1973), and biology is the basis of medicine. Evolutionary medicine takes the view that contemporary diseases are related to incompatibility between the environment in which humans currently live and the conditions under which human biology evolved and genomes have been shaped by a different environment during biological evolution. Human activity has recently acutely modified the environmental conditions in which all living beings live. The main result is a spectacular increased lifespan and pronounced change in the medical landscape. The concept of evolutionary medicine applies to every disease state. From an epidemiological viewpoint, the simultaneous increased incidence in both autoimmune (type 1 diabetes, Crohn's disease, etc.) and allergic (asthma, childhood allergy, etc.) diseases is inversely related to the drop in infectious diseases. The so-called hygiene hypothesis is now solidly established and its mechanism is well documented. It involved both a strong genetic component and a dysregulation of the immune system with a main role attributed to interleukin-10. The conflict resulting from the free availability of food and salt and the remaining fat- and salt-retaining genes is now considered as a major determinant of the rising epidemic incidence of obesity, arterial hypertension and type 2 diabetes. The so-called metabolic syndrome is a summary of these different components and represents a major goal for contemporary preventive medicine. For example, genome-wide association studies have currently identified at least nine genes which are significantly associated with type 2 diabetes. Cancer may be viewed as an accelerated form of evolution at the level of cancer cells. About one third of the hundreds of mutations which have been identified so

---

B. Swynghedauw
Centre de Recherches Cardiovasculaires de l'INSERM, Hôpital Lariboisière, 41 Bd de la Chapelle, 75475 Paris CEDEX, France
Bernard.Swynghedauw@larib.inserm.fr

far in cancers are subject to an evolutionary pressure with a nearly doubling of the nonsynonymous to synonymous substitutions ratio. The genome–environment relationship is a major paradigm for a biologist. It is equally important in medicine, and diseases can now be ascribed as a threshold on a norm reaction curve that depends on a given genotype.

12.1 Introduction

"Nothing makes sense in biology except in the light of evolution" (Dobzhansky 1973). This aphorism should obviously also apply to medicine, and evolutionary biology has an important role to play in medicine. The environments in which humans currently live have indeed resulted in both the emergence of new diseases and profound modifications of the medical landscape. Evolutionary medicine takes the view that present diseases are related to incompatibility between contemporary lifestyles and environments which have been radically modified by the human activity itself and a genome that was shaped over billions of years by different conditions and had remained unchanged during the rather short duration of humanity.

Evolutionary medicine is not a new concept; pioneers in the field were Nesse (Nesse and Williams 1994), Trevathan et al. (1999) and Stearns (1999). Unfortunately, the concept itself has often been discredited by a flood of ideological, theological and "natural" considerations. In addition, integration of the relevant evolutionary paradigm in medical training is far from being achieved, and in medical faculties this particular chapter of biology is still rarely even evoked in most countries (Meagher 2007; MacCallum 2007). This brief review article aims to revisit the basis of an evolutionary approach of the medical paradigm in 2008. It only covers a limited portion of the topic and concentrates on (1) the disease condition viewed as a reaction norm and (2) a few examples of medical applications.

12.2 Illness, a Threshold on a "Reaction Norm"

Phenotypic plasticity is a developmental and evolutionary phenomenon and plastic responses to temperature, predators and photoperiod are good examples of factors influencing fitness, but there are other examples in which the adaptational nature of plasticity is far from being evident (DeWitt and Scheiner 2004). There is a genetic basis to plastic traits with some aspects of environmental response being more genetically malleable than are other aspects and many studies on phenotypic plasticity have identified both plastic genes, i.e., genes that affect the phenotype response to the environment, and allele sensitivity, i.e., the differential expression of genes to different environments (reviewed in Doughty and Reznick 2004). The concept of "reaction norm" coined in 1909 by Woltereck (1909) has been proposed to describe and model the phenotype–environment relationships for various genotypes.

The adaptational nature of the reaction norms is still a debatable issue, but this concept is now widely accepted and even modelized by several groups of evolutionary biologists. Interestingly, the phenotype–environment curve is generally nonlinear and could be either exponential or bell-shaped (DeWitt and Scheiner 2004).

To apply reaction norm to medicine, the disease condition has to be considered as the threshold which occurs on a phenotype–environment curve for a given genotype (Fig. 12.1). A clinical manifestation is a trait, a disease condition is a phenotype and a patient is a witness to the plasticity of his phenotype, even if this plasticity is maladaptative.[1] A disease condition would generally result from a conflict between a genome and the environment, but the respective roles of the two components vary with two opposite situations, namely, severe monogenic diseases such as hemophilia or cystic fibrosis, in which genetics plays nearly exclusively the main role, and bone

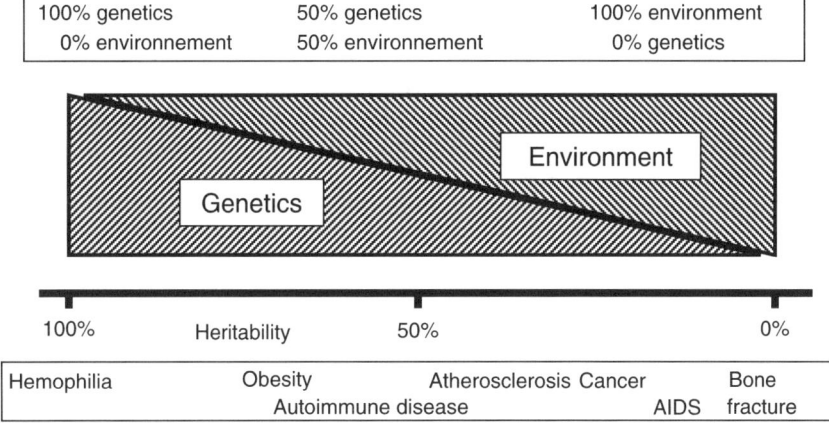

Fig. 12.1 The two partners in disease conditions: genetics (or nature) and environment (or nurture). There are rare diseases which have a 100% genetic basis, such as most monogenic diseases. Hemophilia is the best known, but sickle cell disease is another condition whose causal mutation is well documented. Diseases which are 100% of environmental origin are extremely common, and include every trauma, bleeding and accidents. The most common diseases are usually multifactorial with a variable hereditary component and a variable level of heritability, but several infections, such as AIDS or tuberculosis, can have a hereditary component; for example, people with a rare mutation on an immune component are, in part, protected against infections with the Koch bacillus

---

[1] The situation is not so simple in medicine, and there are examples of genetic mutations that both protect individuals and are responsible for severe biological deficit. The most classical examples are mutations on hemoglobin molecule, the so-called hemoglobinopathies, that both protect the bearers against malaria and provoke deadly anemia. Then, there is an interesting evolutionary conflict which is highly dependent on the environment in countries such as sub-Saharian African countries in which the average lifespan is around 40 years. In these countries, the risk linked to anemia is much less important that the risk due to malaria because in heterozyotous bearers of the mutation, anemia occurs late in life, while malaria may affect and kill infants. The mutation is finally protective and adaptational (discussed in detail in Jobling et al. 2004).

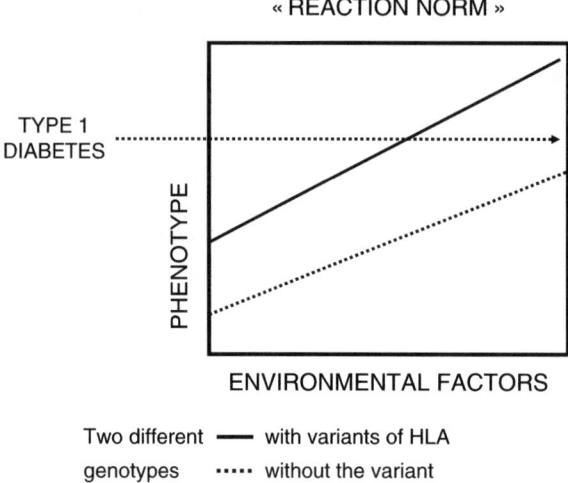

Fig. 12.2 A typical reaction norm applied to type 1 diabetes. Type 1 diabetes is an autoimmune disease which happens in people with several susceptibility genes. Amongst these genes are the genes belonging to the human leukocyte antigen (HLA) system, a major component of the immune system. People with the appropriate variant of the HLA system develop diabetes for a given level of a certain environmental factor. Roughly speaking, the environmental factors involved are the reduced incidence of the infections

fracture, in which the environment is a unique determinant (although that there also genetics determinants that may cause or facilitate a bone fracture) (Fig. 12.2). In practice, intermediary situations account for the majority of disease conditions, with genetics representing from 10% of the determinants (there are cases of familial tuberculosis caused by specific inherited nonfunctional variants of *NRAMP1*; Bellamy et al. 1998) up to 60% (the heritability of obesity, for example, is around 65%). The main objective of present research on common diseases such as diabetes, hypertension, and cancer is to decipher such a relationships, and these investigations basically consist in determining a threshold at which the disease traits appear for a given genotype (Swynghedauw 2006; Wright and Hastie 2007).

12.3 The Central Mechanism at the Base of Evolutionary Medicine

There are at the least three different domains in which our present genome should be in conflict with an environment recently modified by anthropogenic influences, namely, the drop in infections and inflammation, the free availability of food and salt and the consecutive rise in obesity and arterial hypertension and the multiple medical consequences of global warming.

12.3.1 Infections and Susceptibility to Autoimmune and Allergic Diseases

The decrease in the incidence of infectious diseases of bacterial, viral and parasitic origin is extremely well documented and, in developed countries, is a crucial and primary determinant of the decrease in neonatal mortality and rise in lifespan. Such a change could be attributed to both the large utilization of antibiotics and vaccination, and also to improved hygiene and more generally to the amelioration of economic and educational status.

Nevertheless, in the meantime, epidemiology studies have provided evidence of a steady rise in incidence of both autoimmune and allergic diseases. The autoimmune diseases are disorders where the immune system erroneously targets self-antigens, leading to organ-specific or systemic tissue damage. Such aberrant responses result from a breach in tolerance and cellular or humoral autoreactivity. Allergic and autoimmune diseases share common physiopathological mechanisms. The prevalence of several autoimmune diseases such as type 1 diabetes, Crohn's disease and related inflammatory bowel diseases, multiple sclerosis (which results from autoimmune responses to myelin in the central nervous system) and also of allergic diseases such as asthma, rhinitis and atopic dermatitis began to increase in the 1950s, and the incidence of most of these manifestations has doubled or tripled during the last decade (Bach 2002). There are convincing data on the incidence of the main autoimmune diseases in populations migrating from underdeveloped countries to Western countries, and this is attributed to the level of infections in these respective parts of the world. For example, in the UK, the incidence of type 1 diabetes among children of Pakistanis is the same as in nonimmigrants and is 10 times higher than in Pakistanis living in Pakistan. The same rationale applies to allergic diseases; asthma, for example, which is so common in our countries, is a rare condition in India.

The respective influences of genetic and environmental factors in the pathophysiological mechanisms of autoimmune and allergic diseases began to be better understood. Genome-wide association studies[2] have recently supported the existence of susceptibility genes associated with type 1 diabetes and have evidenced at least a dozen genes strongly and reproducibly associated with this type of diabetes, including *HLA* class genes, *insulin* gene, a *CTLA4* locus, *PTPN22* and the *IFIH1* region, that mostly belong to the immune system (Wellcome Trust Case Control

---

[2] Genome-wide association study is a new and revolutionary approach in medical genetics and is based on (1) the availability of dense genotyping chips made with single-nucleotide polymorphisms (SNPs) covering unequally most of the genome, (2) the growing resources of the International HapMap Consortium (2007), which documents linkage disequilibrium, and is a public resource of common SNPs capturing most of the common genome sequence variability. In the human genome, there are 3.2 billion base pairs and approximately 15 million SNPs and kits using even 500,000 SNPs should cover less than 0.2% of the genome. The second-generation human haplotype map now covers over 3.1 million SNPs, and the statistical association amongst groups of SNPs within haplotype blocks suggests that the identification of a few of the SNPs within haplotype blocks can unambiguously cover more than 90% of the genome.

Consortium 2007). Numerous studies have identified the absence of infections as an environmental determinant of autoimmune diseases, such as, for example, the protection afforded by helminths or childhood growth on a farm. Experimental models have confirmed such a negative correlation, and, for example, the incidence of this type of diabetes increases immediately after breeding conditions is strikingly modified by a specific pathogen-free environment.

Several mechanisms have been proposed to explain such a relationships. The basis of the so-called hygiene hypothesis is unlikely to lie in a conflict between the Th1 and Th2 immune cells and is more likely to involve the production of compounds playing a counter-regulator role. The most documented of these compounds is an anti-inflammatory cytokine, interleukin-10, but there are several others such as nitric oxide and transforming growth factor β. Elevation of interleukin-10 during long-term helminth infections is inversely correlated with allergy (Wills-Kasp 2001; Yazdanbakhsh et al. 2002).

12.3.2 Obesity, Type 2 Diabetes, Arterial Hypertension and the Thrifty Gene Hypothesis

The rising incidence of obesity and type 2 diabetes[3] is one of the major epidemiological challenge of this century and is generally considered as the most salient feature of evolutionary medicine. The so-called thrifty genes hypothesis has many different facets, but it mainly postulates that new nutritional patterns and dietary factors were being played out against a genomic inheritance of coadapted gene complexes with multiple epistasic, multifactorial and pleiotropic relationships. Our current genome structure is the product of billions of years of selective pressure under a mobile, pretty hectic, primarily plant-eating but omnivorous ecological niche with finally a short life expectancy. In addition our ancestors had to be adapted to exercising. The bearers of such a genome are now exposed to sedentary lifestyles, and fat- and salt-rich, but fiber-poor diets with a multitude of genes or variants which had been previously selected for a totally opposite purpose (Neel 1962; Trevathan 1999). Obesity and its major deleterious consequences, namely, type 2 diabetes arterial hypertension, hyperlipidemias, clinical manifestations of atherosclerosis[4] and cancer, are the

---

[3] Diabetes is an extremely frequent disease in our country with a strong heritability. There are two types of diabetes. Type 1 is an autoimmune disease which basically destroys the pancreatic insulin-producing cells and is a severe form of diabetes frequently beginning during childhood. By contrast, type 2 happens in adults and is principally linked to overnutrition and obesity. From a pathophysiological point of view, type 2 diabetes is accompanied by high plasma levels of insulin and the primary causal mechanism is insulin resistance.

[4] Atherosclerosis is characterized by a progressive deposit of lipids (cholesterol) in the arterial intima which trigger an inflammatory reaction and fibrosis in the big arteries, resulting in arterial obstruction. This is one of the main causes of morbidity and mortality in Western countries. Atherosclerosis is causally related to both coronary heart diseases (myocardial infarction) and stroke.

results of such a conflict and most of the currently recommended changes in diets and lifestyle are presently based on the so-called Paleolithic diet (Eaton and Konner 1985).

The respective influences of genes and the environment in the pathogenesis of obesity are now starting to be better understood: "the last decade have been the golden age of obesity research" (Flier 2004). Survival is obviously more accurately and more acutely threatened by starvation than by overnutrition, and the system is more robustly organized to stimulate energy intake and fat stores than to expend energy. Then when food becomes rare the different stores of energy, such as adipose tissue, promote survival and the evolutionary process should promote selection of thrifty genes over genes activating intracellular catabolism.

Obesity is a highly heritable condition (the heritability is around 60%) and genetic research, including recent reproducible, large scale approaches based on genome-wide association studies, has evidenced many body mass index associated genes, including *ESRI, PPARG, ADIPOQ, INSIG2, LEP, ESRI, SSTR2* and also *LRP1B, VIP, ADRB1, NPY2R, HSD3B1, ADRA1B, IL6R, AGTR1*, and *FSHR* (Cupples et al. 2007)[5] or with established obesity as *FTO* (Frayling et al. 2007; Boomsma et al. 2002). A plethora of genes that control food intake have been identified so far. In rodents, mutations at these levels cause profound obesity and key signaling efferent or afferent pathways, including the leptin pathway, insulin, gut and stomach signals, have been identified with an integration at the hypothalamic level (Schwartz and Porte 2005). Several pharmacological compounds based on this concept have been discovered and some of them are already commercially available (van Gaal et al. 2005).

The association of obesity with type 2 diabetes, an increased plasma level of lipids, arterial hypertension and clinical manifestations of atherosclerosis (including myocardial infarction and stroke) forms a cluster which is commonly identified as the metabolic syndrome (Eckel et al. 2005). On the basis of experimental data (MacMillen and Robinson 2005) and epidemiology studies (Barker et al. 1989), the metabolic syndrome can now be viewed as a consequence of prenatal stimuli acting through an epigenetic mechanism, including DNA methylation and modifications of genetic imprinting (Junien 2005). From an evolutionary point of view, such a hypothesis (also called the Barker hypothesis) represent a step towards a better understanding, although through a different and unexpected pathway, of the relation between biological evolution and the radical modifications of the present medical status.

12.3.3 The Multiple Medical Consequences of Global Warming

The topic is hot. The anthropogenic origins of the present changes in the climate are well documented: "The IPCC report has served a useful purpose in removing

---

[5] Gene nomenclature can be found at http://www.genecards.org.

the last ground from under the sceptics' feet, leaving them looking marooned and ridiculous" (Anonymous 2007). The consequences of such a sudden event on public health are extremely diverse and may include an evolutionary conflict between our genome, shaped over, at least, 200,000 years by a rather stable environment and the recent modifications of the climate. The conflict is likely caused by the rapidity and the unpredictability of the changes more than by the average increase in global temperature and it involves totally different issues. Global climate change not only increases the average temperature of our planet, with numerous urban heat islands, it also augments the severity and number of heat waves and tropical hurricanes, and has several other diverse consequences in terms of hydratation, dryness, floods, population migrations, etc.

From a medical point of view, there are four types of consequences:

1. The simple elevation of the average external temperature is accompanied by increased global mortality and morbidity, the mortality-external temperature curve is a J-shaped curve, with the warm branch being more pronounced than the cold one (Haines et al. 2006; MacMichael et al. 2006). A recent study of 50 different cities confirmed that global, and more specifically cardiovascular mortalities were enhanced at the two temperature extremes (Medina-Ramón and Schwartz J 2007).
2. Acute heat waves, such as those which happened in France in August 2003, have been studied in detail by several groups. The mortality which was observed during the recent heat waves was not compensated by harvesting, strongly suggesting that we were dealing with heat stroke, and that such an increased mortality was reflecting more the limits of our adaptational capacities than aggravation of a previously altered health status (Le Tertre et al. 2006).
3. Climate changes have modified the repartition and virulence of pathogenic agents (dengue, malaria, etc.) and above all their vectors (Hales and Woodward 2003; Hopp and Foley 2003). Such modifications were exponential and are likely to reflect the biological properties of parasites (Patz and Olson 2005).
4. Indirect consequences of global warming include variations in the hydraulic cycle, the new form of tropical hurricanes and many different changes affecting both biodiversity and ecosystems. They will likely result in an increased level of poverty (Haines et al. 2006).

These finding gave rise to several basic biological questions, rarely evoked, and that concern the limits of the adaptational capacities of the human genome. Our genome has indeed been shaped in the past by a rather cold environment which has been acutely modified. The immediate physiological regulation includes sweating and skin vasodilatation. The latter may strongly enhance cardiac output, which explains heat-induced cardiac decompensation. Long-term regulation depends upon the numerous mechanisms of uncoupling of mitochondrial respiration. For the moment, the thermolytic mechanisms and their regulation are rather poorly documented (Silva 2006). Nevertheless this emphasizes the potential role of the mitochondrial genome in adaptation.

12.4 Medical Applications, Carcinogenesis

The acute modifications of our current environment have deeply modified the medical landscape. At the moment, the first major event is the impressive increase in lifespan, at least in developed countries or regions.[6] The average lifespan in our countries in 2007 was around 76 years, whereas it was 30–40 years in the eighteenth century. As a consequence, most patients seen by general practitioners are elderly and suffer from diseases which had never been or were rarely experienced before, such as cancer and the clinical manifestations of atherosclerosis. Such a rise in the incidence of new diseases reflects a complicated interplay between (1) low-penetrance genes, (2) a more prolonged contact with several risk factors, such as plasma glucose and cholesterol, pollution, overnutrition and tobacco smoking and (3) the specific cellular effects of senescence which favors both cancer (Finkel et al. 2007) and atherosclerosis.

Cancer may be an exemplary application of evolutionary medicine. Are neoplasms microcosms of evolution (Merlo et al. 2006)? Is cancer the Darwinian downside of past success (Greaves 2002)? At first glance and from an evolutionary perspective cancer may be viewed as a genetically and epigenetically heterogeneous population of individual cells (Hanahan and Weinberg 2000), each mutation in cancer cells conferring to these cells an advantage, even though the mutation is detrimental for the organism. Assembling the repertoire of the numerous mutations present in cancers is currently in progress. The first works have already identified more than 1,000 mutations in several cancers. In cancer, mutations happen randomly: some of them are located in germinal lines (and make a minority of cancers heritable), but most of them are on autosomal cells. A recent study has, for example, identified more than a thousand mutations in a group of cancers of different origins; most of them were located on the coding part of tyrosine kinases and on several genes involved in reparative process. Based on the dN/dS ratio, 150 of these mutations, out of 921, were identified as "drivers genes" (Greenman et al. 2007). Such an approach allowed proteins to be selected which may play an important role in carcinogenesis, and should constitute specific targets for therapy.

Inflammation is an important determinant of both the development and the severity of the disease process (Cousssens and Werb 2002). Many cancers arise from sites of infections or chronic irritations, and it is also clear that the tumor environment is orchestrated by inflammatory cells which are indispensable participants of cell proliferation and migration. Inflammatory cascade is one of the tools that tumor cells have utilized for migration and development. Another important component of evolutionary medicine, namely, obesity, and overnutrition are known to be risk factors for cancer, and more specifically for breast cancer. A molecular basis for this linkage may lie in the inflammatory reaction which is present in adipose tissue (Lorincz and Sukumai 2006).

---

[6] There are still important differences, mostly linked to economic status, throughout the world both in mortality rates and in the causes of death (Yusuf et al. 2001).

References

Anonymous (2007) Light at the end of the tunnel. Nature 445:567
Bach JF (2002) The effect of infections on susceptibility to autoimmune and allergic diseases. N Engl J Med 347:911–920
Barker DJ, Winter PD, Osmond C, Margetts B, Simmonds SJ (1989) Weight in infancy and death from ischemic heart disease. Lancet 2:577–580
Bellamy R, Ruwende C, Corrah T, McAdam KP, Whittle HC, Hill AV (1998) Variations in the NRAMP1 gene and susceptibility to tuberculosis in West Africans. N Engl J Med 338:640–644
Boomsma D, Bujsjahn A, Peltonen L (2002) Classical twin studies and beyond. Nat Rev Genet 3:872–882
Coussens LM, Werb Z (2002) Inflammation and cancer. Nature 420:860–867
Cupples LA, Arruda HT, Benjamin EJ, D'Agostino RB Sr, Demissie S, DeStefano AL, Dupuis J, Falls KM, Fox CS, Gottlieb DJ, Govindaraju DR, Guo CY, Heard-Costa NL, Hwang SJ, Kathiresan S, Kiel DP, Laramie JM, Larson MG, Levy D, Liu CY, Lunetta KL, Mailman MD, Manning AK, Meigs JB, Murabito JM, Newton-Cheh C, O'Connor GT, O'Donnell CJ, Pandey M, Seshadri S, Vasan RS, Wang ZY, Wilk JB, Wolf PA, Yang Q, Atwood LD (2007) The Framingham Heart Study 100K SNP genome-wide association study resource: overview of 17 phenotype working group reports. BMC Med Genet 8(Suppl I):SI
DeWitt TJ, Scheiner SM (2004) Phenotypic plasticity. Functional and conceptual approaches. Oxford University Press, New York
Dobzhansky T (1973) Nothing in biology makes sense except in the light of evolution. Am Biol Teach 35:125–129
Doughty P, Reznick DN (2004) Patterns and analysis of adaptative phenotypic plasticity in animals. In: DeWitt TJ, Scheiner SM (eds) Phenotypic plasticity. Functional and conceptual approaches. Oxford University Press, New York
Eaton SB, Konner M (1985) Paleolithic nutrition. A consideration of its nature and current implications. N Engl J Med 312:283–289
Eckel RH, Grundy SM, Zimmet PZ (2005) The metabolic syndrome. Lancet 365:1415–1428
Finkel T, Serrano M, Blasco MA (2007) The common biology of cancer and ageing. Nature 448:767–774
Flier JS (2004). Obesity wars: molecular progress confronts an expanding epidemic. Cell 116:337–350
Frayling TM, Timpson NJ, Weedon MN, Zeggini E, Freathy RM, Lindgren CM, Perry JR, Elliott KS, Lango H, Rayner NW, Shields B, Harries LW, Barrett JC, Ellard S, Groves CJ, Knight B, Patch AM, Ness AR, Ebrahim S, Lawlor DA, Ring SM, Ben-Shlomo Y, Jarvelin MR, Sovio U, Bennett AJ, Melzer D, Ferrucci L, Loos RJ, Barroso I, Wareham NJ, Karpe F, Owen KR, Cardon LR, Walker M, Hitman GA, Palmer CN, Doney AS, Morris AD, Smith GD, Hattersley AT, McCarthy MI (2007) A common variant in the FTO gene is associated with body mass index and predisposes to childhood and adult obesity. Science 316:889–892
Greaves M (2002) Cancer causation: the Darwinian downside of past success? Lancet Oncol. 3:244–251
Greenman C, Stephens P, Smith R, Dalgliesh GL, Hunter C, Bignell G, Davies H, Teague J, Butler A, Stevens C, Edkins S, O'Meara S, Vastrik I, Schmidt EE, Avis T, Barthorpe S, Bhamra G, Buck G, Choudhury B, Clements J, Cole J, Dicks E, Forbes S, Gray K, Halliday K, Harrison R, Hills K, Hinton J, Jenkinson A, Jones D, Menzies A, Mironenko T, Perry J, Raine K, Richardson D, Shepherd R, Small A, Tofts C, Varian J, Webb T, West S, Widaa S, Yates A, Cahill DP, Louis DN, Goldstraw P, Nicholson AG, Brasseur F, Looijenga L, Weber BL, Chiew YE, DeFazio A, Greaves MF, Green AR, Campbell P, Birney E, Easton DF, Chenevix-Trench G, Tan MH, Khoo SK, Teh BT, Yuen ST, Leung SY, Wooster R, Futreal PA, Stratton MR (2007) Patterns of somatic mutation in human cancer genomes. Nature 446:153–158
Haines A, Kovats RS, Campbell-Lendrum D, Corvalon C (2006) Climate change and human health: impacts, vulnerability, and mitigation. Lancet 367:2101–2109
Hales S, Woodward A (2003) Climate change will increase demands on malaria control in Africa. Lancet 363:1775

Hanahan D, Weinberg RA (2000) The hallmarks of cancer. Cell 100:57–70

Hopp MJ, Foley JA (2003) Worldwide fluctuations in dengue fever cases related to climate variability. Clim Res 25:85–94

International HapMap Consortium (2005) A haplotype map of the human genome. Nature 437:1299–1320

Jobling MA, Hurles ME, Tyler-Smith C (2004) Human evolutionary genetics. Origins, peoples and disease. Garland, New York

Junien C, Gallou-Kabani C, Vigé A, Gross M-S (2005) Epigénomique nutritionnelle du syndrome métabolique. Med Sci (Paris) 21:396–404

Le Tertre A, Lefranc A, Eilstein D, Declercq C, Medina S, Blanchard M, Chardon B, Fabre P, Filleul L, Jusot JF, Pascal L, Prouvost H, Cassadou S, Ledrans M (2006) Impact of the 2003 heatwave on all-cause mortality in 9 French cities. Epidemiology 17:75–79

Lorincz AM, Sukumai S (2006) Molecular links between obesity and breast cancer. Endocr Relat Cancer 13:279–292

MacCallum CJ (2007) Does medicine without evolution make sense? PLoS Biol 5:e112–e114

MacMichael AJ, Woodruff RE, Hales S (2006) Climate change and human health present and future risks. Lancet 367:859–869

MacMillen C, Robinson JS (2005) Developmental origins of the metabolic syndrome: prediction, plasticity, and programming. Physiol Rev 85:571–633

Meagher TR (2007) Is evolutionary biology strategic science? Evolution Jan 239–244

Medina-Ramón M, Schwartz J (2007) Temperature, temperature extremes, and mortality: a study of acclimatisation and effect modification in 50 US cities. Occup Environ Med 64:827–833. doi:10.1136/oem.2007.033175

Merlo LM, Pepper JW, Reid BJ, Maley CC (2006) Cancer as an evolutionary and ecological process. Nat Rev Cancer 6:924–935

Neel JV (1962) Diabetes mellitus: a "thrifty" genotype rendered detrimental by "progress"? Am J Hum Genet 14:353–362

Nesse RM, Williams G (1994) Why we get sick: the new science of Darwinian medicine. Times Books, New York

Patz JA, Olson SH (2005) Malaria risk and temperature: influences from global climate change and local land use practices. Proc Natl Acad Sci USA 103:5635–5636

Schwartz MW, Porte D Jr (2005) Diabetes, obesity, and the brain. Science 307:375–379

Silva JE (2006) Thermogenic mechanisms and their hormonal regulation. Physiol Rev 86:435–464

Stearns SC (1999) Evolution in health and disease. Oxford University Press, Oxford

Swynghedauw B (2006) Phenotypic plasticity of adult myocardium. Molecular mechanisms. J Exp Biol 209:2320–2327

Trevathan WR, Smith EO, McKenna JJ (1999) Evolutionary medicine. Oxford University Press, Oxford

van Gaal LF, Rissanen AM, Scheen AJ, Ziegler O, Rössner S, RIO-Europe Study Group (2005) Effects of the cannabinoid-1 receptor blocker rimonabant on weight reduction and cardiovascular risk factors in overweight patients: 1-year experience from the RIO-Europe study. Lancet 365:1389–1397

Wellcome Trust Case Control Consortium (2007) Genome-wide association study of 14,000 cases of seven common diseases and 3,000 shared controls. Nature 447:661–678

Wills-Karp M, Santeliz J, Karp CL (2001) The germless theory of allergic disease: revisiting the hygiene hypothesis. Nat Rev Immunol 1:69–75

Woltereck R (1909) Weitere experimentelle Untersuchungen über Artveränderung, spezielle über das wesen quantitativer Artunterschiede bei Saphniden. Verh Dtsch Zool Ges 19:110–172

Wright A, Hastie N (2007) Genes and common diseases. Genetic in modern medicine. Cambridge University Press, Cambridge

Yazdanbakhsh M, Kremaner PG, van Ree R (2002) Allergy, parasites, and the hygiene hypothesis. Science 296:490–494

Yusuf S, Reddy S, Ounpuu S, Anand S (2001) Global burden of cardiovascular diseases. Part II: variations in cardiovascular disease by specific ethnic groups and geographic regions and prevention strategies. Circulation 104:2855–2864

Chapter 13
An Overview of Evolutionary Biology Concepts for Functional Annotation: Advances and Challenges

Anthony Levasseur and Pierre Pontarotti

Abstract Numerous genome sequencing projects have yielded more and more data to help analyze and gain a better understanding of genetic diversity in the living world. Genome annotation enables us to decipher raw data and identify protein-coding genes and their function. After structural annotation, the next step is predicting protein function (functional annotation). All the genes currently available in the sequenced genomes will never be studied experimentally, and so the most reliable and accurate theoretical approaches need to be considered for function prediction. Different approaches have already been developed for functional annotation. Here we emphasize the use of evolutionary biology concepts as an improved and sensitive method for predicting protein function. In the future, functional annotation based on evolutionary biology could be included in a more general approach to study the impact of the environment on the genome at the community scale.

13.1 Improving Functional Annotation Using Evolutionary Biology

Typically, functional annotation uses sequence-based methods to make direct sequence-to-sequence comparison and predict a putative function to match the best database hit. This approach, based on the search for similarity, implies that the proteins can be homologous, i.e., descended from a common ancestral gene. There are two different types of homologous genes; orthologs and paralogs. They evolve by vertical descent from a single ancestral gene (speciation event) or by duplication, respectively. In evolutionary history, homologous genes undergo different events, such as mutations, leading in some cases to a functional shift. These functional

---

A. Levasseur and P. Pontarotti
Laboratoire Evolution Biologique et Modélisation, Case 19, UMR 6632,
Université d'Aix-Marseille/CNRS, 3 Place Victor Hugo, 13331 Marseille CEDEX 03, France
anthony.levasseur@univ-provence.fr; pierre.pontarotti@univ-provence.fr

shifts (or co-option events) are the result of a protein modification at the sequence level (e.g., changes of active-site or ligand recognition) or at the transcriptional level (e.g., changes in the expression pattern) (Ganfornina and Sanchez 1999). The distinction between orthologs and paralogs is critical for reliable functional annotation. It is generally assumed that one-to-one orthologs are functionally equivalent. This is not so for the paralogs, since after a duplication event the two duplicates can be subfunctionalized, or one of the copies can be lost or evolve without functional constraint towards some new function (neofunctionalization). Paralogy is therefore closely associated with functional diversification and specialization. These important events can be deciphered by a phylogenetic study, and information can be integrated for functional annotation. By superimposition of the gene tree and the function tree, some nodes can be annotated for a specific function and hold throughout the lineages. Compared with the methods based on the similarity search, this approach, based on evolutionary biology concepts, is more informative and can enhance the accuracy and reliability of the functional inference. Software for this purpose (based on a Bayesian approach) was developed by Engelhardt et al (2005). It makes it possible to improve functional prediction. A software platform (FIGENIX) has been developed that provides a straightforward, accelerated phylogenetic tree construction using automatic pipelines and propagates the annotation from one ortholog to another (Gouret et al. 2005).

In addition to the orthology/paralogy relationship, functional annotation can be further enhanced by integrating further information from modern evolutionary biology concepts (Gaucher et al. 2002; Sjolander 2004). Other essential information can be automatically added to the phylogenomic approach to extend the threshold of the functional prediction. Some additional methods of interest are proposed in the next section.

13.2 Considering the Evolutionary Shift

The neutral theory of evolution states that the sites of greatest functional significance come under the strongest selective constraint (Fitch and Markowitz 1970). This implicitly means that functional constraint among amino acids changes when a change in protein function (functional shift) occurs during evolutionary history. One way to improve functional annotation would be automatically and consistently to identify functional shifts in the phylogenetic tree, and their effect on the subsequent functional annotation. Thus, in addition to the well-known phylogenomic approach, systematic integration of the functional shift in new software could be valuable for functional annotation. For instance, functional shift could be confined to a specific branch, and so the modification of function could be inferred for all the descending lineages (Fig. 13.1). Functional shift occurs by relaxed functional constraint or by positive selection, and can be detected by computational methods that take into account the patterns of replacement. Currently, original approaches have emerged that consider (1) amino acid replacements (nonsynonymous substitution) alone or (2) the ratio of nonsynonymous to synonymous substitution.

13 An Overview of Evolutionary Biology Concepts for Functional Annotation

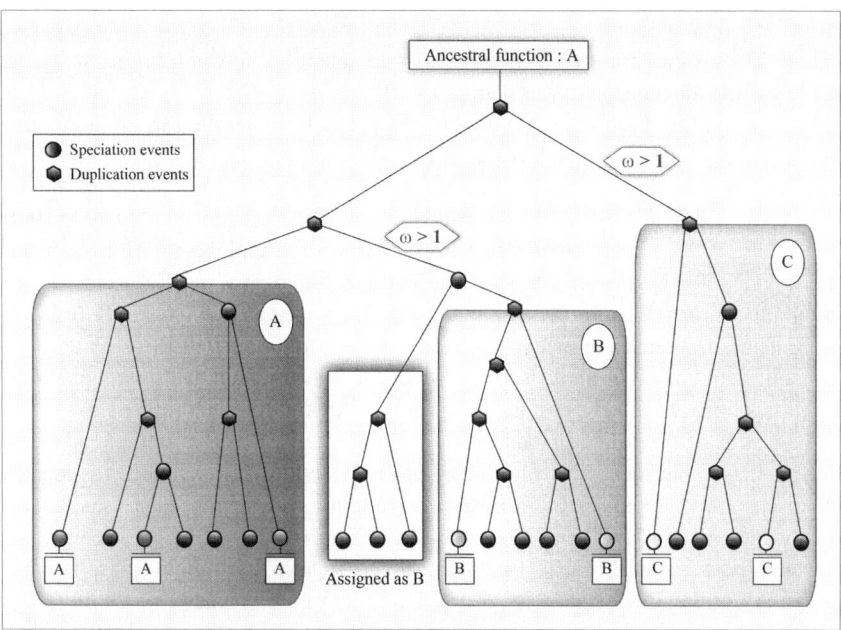

Fig. 13.1 Functional annotation based on evolutionary biology concepts. On the basis of experimental data and phylogenetic history, nodes and their corresponding leaves can be functionally annotated as *A*, *B* or *C*. The function of unknown proteins or new members can be predicted according to their evolutionary history and membership of particular lineages. Also, detection of evolutionary shift ($\omega > 1$) along a specific branch may make it possible to predict a functional divergence in the descent lineage. *Gray circles* experimental functional data, *black circles* functional prediction. All nonlabeled branches have $\omega < 1$

The method based on amino acid replacements uses two types of gamma models that consider how the evolutionary rates of amino acid replacement differ among sites in a protein sequence. In the homogeneous gamma model, rapid, moderate and slow positions conserved their respective rates throughout the evolutionary tree (Yang 1996). However, in the case of functional shift, evolutionary rates of sites will be different along the branches of the tree (heterotachy) owing to altered functional constraints. Accordingly, a nonhomogeneous gamma model allowing a slow site to become a fast one (or vice versa) was developed. This model can thereby locate sites that are probably involved in functional divergence (potential targets for site-directed mutagenesis).

Another informative approach for functional annotation consists in detecting positive selection using a comparison of the relative ratio of nonsynonymous to synonymous substitutions (Miyata and Yasunaga 1980). The nonsynonymous to synonymous substitution rate ratio (denoted ω) is used as a measure of selective pressure at the protein level. A value of ω greater than 1 indicates that nonsynonymous substitutions offer an advantage and are set at a higher rate than synonymous ones, whereas $\omega = 1$ and $\omega < 1$ indicate neutral evolution or purifying selection, respectively. Most amino acids are under strong structural and functional constraints, and positive selection usually acts on a few sites for a short period of time. Thus, detection of significant positive selection between homologous sequences is usually unsuccessful if ω is calculated over all sequences and over all the time separating them. A great deal of effort has been devoted to implementing models and testing positive selection on an individual branch (or a set of branches) of a phylogenetic tree (branch methods) or on individual codon sites (site methods) (Yu and Irwin 1996; Messier and Stewart 1997; Zhang et al. 1997, 1998, 2005; Yang 1998; Nielsen and Yang 1998; Suzuki and Gojobori 1999; Yang and Bielawski 2000). Recently, a branch-site method for testing positive selection on individual codons along specific lineages was developed (Yang and Nielsen 2002). The branch-site method considers branches in the phylogenetic tree as foreground and background lineages and constructs a likelihood ratio test by comparing a model that allows positive selection in the foreground with a model that refuses it.

The power of the functional annotation can be improved by systematically considering these methods of evolutionary shift detection. For instance, a specific branch can be targeted during a period of functional divergence on the basis of the ω value and the functional inference can be improved (Fig. 13.1). Recently, we correlated positive selection to functional divergence in the fungal lipase/feruloyl esterase A family (Levasseur et al. 2006). This example illustrates how evolutionary shift can be used for functional annotation as the branch leading to the functional diversification evolved under positive selection. However, the use of positive selection is not strictly an indicator of functional shift per se and must be carefully integrated with other information. Few published examples of relaxed or positive selection have been linked to actual functional shifts (Levasseur et al. 2007).

Another interesting variable to be taken into account is the relation between branch length and expression pattern. Previous work hypothesized that long branches were correlated with specific and restricted gene expression. By contrast, genes with short branches seem to be extensively or ubiquitously expressed (Paillisson et al. 2006; Balandraud et al. 2005). A detailed, exhaustive study is still required to evaluate the general trend of this correlation. This information could potentially be exploited for functional annotation.

13.3 Evolutionary Biology Concepts in the Genomic Era

To extend the scope of analysis, concepts should be considered at a level higher than that of a single gene, i.e., at a community genomic scale. Functional annotation based on evolutionary biology would be much more informative applied to a total gene data set from complete genomes. This genomic approach makes it possible to investigate specific features of genome evolution in greater depth and can connect evolutionary events with specific aspects of physiological function or environment (especially extreme environment). Hence, the main step consists in placing evolutionary history in its environmental context. A better understanding of the role of the environment in genome evolution could be achieved by studying overall gene diversity through a comparative genomic approach between organisms or more generally at the community level ("the community as a whole organism").

13.3.1 Comparative Genomic Approach

The comparative genomic approach through various genomes of different biotopes is an interesting method to study genome/environment relationships. Functional annotation at the genome scale enables us to extend our knowledge of the role of the environment in genome evolution. For instance, a growing number of fungal genomes are currently available, and comparative genomic approaches should advance our knowledge of the mechanisms of adaptation whereby fungi persist in the environment (Galagan et al. 2005). Fundamental questions could be addressed, for instance: How can we explain the enzymatic divergence between white rot, brown and mold fungi in the light of genome evolution with a special focus on the genic repertory discrepancies? Identifying the part of the genome content that has evolved under positive selection would be highly relevant to the understanding of the functional repertories of different fungi, and could be correlated to their lifestyle and degrader behavior.

13.3.2 Towards a Functional Annotation on the Community Scale

In addition to the genome-to-genome comparison, a large-scale study could be achieved by a metagenomic approach. The genomic repertoire analysis of different communities from various biotopes is an informative means to unravel the genome/environment relation. Metagenomic approaches enable us to bypass technical bottlenecks such as the difficulties of microorganism cultivation and the reproducibility of natural conditions in an artificial environment. For instance, the distributional patterns of genes from planktonic microbial communities from the ocean surface to near the sea floor allows the identification of depth-variable trends in gene content and metabolic pathway components (DeLong et al. 2006). Combined with a competitive functional annotation, a metagenomic approach based on microbial communities from different biotopes (e.g., from various geographic samplings or from extremely polluted sites) will offer new ways to evaluate the impact of environment on genome content.

13.4 Conclusion

Today, evolutionary approaches can advance functional annotation, but they are not yet routinely used. The integration of evolutionary biology concepts for functional annotation leads to a better inference and a better understanding of the "functional plasticity" of proteins. Considering branch lengths and positive selection, new parameters could be included to balance functional prediction and decipher evolutionary events that shape the history of genes. In addition to the gene-to-gene analysis, the integration of evolutionary biology concepts for global genome comparisons will be a fundamental step in elucidating the relationship between organisms and their environment. In the future these evolutionary concepts could be integrated into competitive software dedicated to functional annotation.

Acknowledgements A.S.G.L. thanks the University of the Mediterranean (Aix-Marseille 2) for his postdoctoral fellowship.

References

Balandraud N, Gouret P, Danchin EG, Blanc M, Zinn D, Roudier J, Pontarotti P (2005) A rigorous method for multigenic families functional annotation: the peptidyl arginine deiminase (PADs) proteins family example. BMC Genomics 6:153

DeLong EF, Preston CM, Mincer T, Rich V, Hallam SJ, Frigaard NU, Martinez A, Sullivan MB, Edwards R, Brito BR, Chisholm SW, Karl DM (2006) Community genomics among stratified microbial assemblages in the ocean's interior. Science 311:496–503

Engelhardt BE, Jordan MI, Muratore KE, Brenner SE (2005) Protein molecular function prediction by Bayesian phylogenomics. PLoS Comput Biol 1:432–445

Fitch WM, Markowitz E (1970) An improved method for determining codon variability in a gene and its application to the rate of fixation of mutations in evolution. Biochem Genet 4:579–593

Galagan JE, Henn MR, Ma LJ, Cuomo CA, Birren B (2005) Genomics of the fungal kingdom: insights into eukaryotic biology. Genome Res 15:1620–1631

Ganfornina MD, Sanchez D (1999) Generation of evolutionary novelty by functional shift. BioEssays 21:432–439

Gaucher EA, Gu X, Miyamoto MM, Benner SA (2002) Predicting functional divergence in protein evolution by site-specific rate shifts. Trends Biochem Sci 27:315–321

Gouret P, Vitiello V, Balandraud N, Gilles A, Pontarotti P, Danchin EG (2005) FIGENIX: intelligent automation of genomic annotation: expertise integration in a new software platform. BMC Bioinformatics 6:198

Levasseur A, Gouret P, Lesage-Meessen L, Asther M, Asther M, Record E, Pontarotti P (2006) Tracking the connection between evolutionary and functional shifts using the fungal lipase/ feruloyl esterase A family. BMC Evol Biol 6:92

Levasseur A, Orlando L, Bailly X, Milinkovitch MC, Danchin EG, Pontarotti P (2007) Conceptual bases for quantifying the role of the environment on gene evolution: the participation of positive selection and neutral evolution. Biol Rev Camb Philos Soc. 82:551–572

Messier W, Stewart CB (1997) Episodic adaptive evolution of primate lysozymes. Nature 385: 151–154

Miyata T, Yasunaga T (1980) Molecular evolution of mRNA: a method for estimating evolutionary rates of synonymous and amino acid substitutions from homologous nucleotide sequences and its application. J Mol Evol 16:23–36

Nielsen R, Yang Z (1998) Likelihood models for detecting positively selected amino acid sites and applications to the HIV-1 envelope gene. Genetics 148:929–936

Paillisson A, Levasseur A, Gouret P, Callebaut I, Bontoux M, Pontarotti P, Monget P (2007) Bromodomain testis-specific protein is expressed in mouse oocyte and evolves faster than its ubiquitously expressed paralogs BRD2, −3, and −4. Genomics 89:215–223

Sjolander K (2004) Phylogenomic inference of protein molecular function: advances and challenges. Bioinformatics 20:170–179

Suzuki Y, Gojobori T (1999) A method for detecting positive selection at single amino acid sites. Mol Biol Evol 16:1315–1328

Yang Z (1998) Likelihood ratio tests for detecting positive selection and application to primate lysozyme evolution. Mol Biol Evol 15:568–573

Yang Z, Bielawski JP (2000) Statistical methods for detecting molecular adaptation. Trends Ecol Evol 15:496–503

Yang Z, Nielsen R (2002) Codon-substitution models for detecting molecular adaptation at individual sites along specific lineages. Mol Biol Evol 19:908–909

Yang ZH (1996) Among-site rate variation and its impact on phylogenetic analyses. Trends Ecol Evol 11:367–372

Yu M, Irwin DM (1996) Evolution of stomach lysozyme: the pig lysozyme gene. Mol Phylogenet Evol 5:298–308

Zhang J, Kumar S, Nei M (1997) Small-sample tests of episodic adaptive evolution: a case study of primate lysozymes. Mol Biol Evol 14:1335–1338

Zhang J, Rosenberg HF, Nei M (1998) Positive Darwinian selection after gene duplication in primate ribonuclease genes. Proc Natl Acad Sci USA 95:3708–3713

Zhang J, Nielsen R, Yang Z (2005) Evaluation of an improved branch-site likelihood method for detecting positive selection at the molecular level. Mol Biol Evol 22:2472–2479

Index

μ mutations, 3
  beneficial, 3–6, 8, 9, 16
  deleterious, 3, 6, 9, 14, 16

Aboral pore, 127–130
actin, 145
Adiantum, 165, 168, 174, 175
Albert von Kölliker, 96
Albostrians, 176
Allergic Diseases, 201
allopatric fragmentation, 190
allopolyploid, 161, 162
Amborellaceae, 170, 171
Amino acids, 165, 167–169, 173–176
  acid, 168
  aromatic, 168
  basic, 168
  nonpolar, 168
  special, 168
Among-site rate heterogeneity, 53
Anal pore, 127
Angiosperms, 167, 168, 170, 171
anterior chamber, 141, 146, 147
Anthoceros, 168
Anthozoa, 135
Arabidopsis, 175
archetype of the vertebrates, 105
Aromorphoses, 84, 89
Arterial Hypertension, 202
atherosclerosis, 205
atpB, 167, 170
atpF, 176
atrichous isorhiza, 137
autoimmune, 200
Autoimmune Diseases, 201

Barley, 175, 176
Basal disc, 117, 128–130
battery cell, 138
Bayesian Markov Chain Monte Carlo (BMCMC), 29, 37
beetle, 182
Bouin's fixative, 147
Bourgeois society, 87, 88, 90
Branch Length, 155, 161, 162
Bruniaceae, 168

Caenorhabditis elegans, 128, 131
Canal systems, in Foraminifera, 83
canalization, 100
Cancer, 61, 62, 205
  embryonality, 62, 63, 69
  embryonal theory, 61, 62
  stem cells, 61–63, 68
Cancer cells, 61
capsule, 137–142, 144–147
Ceratophyllales, 170
Chaetognaths, 155, 156, 158, 160–162
Chloranthaceae, 170
Chloranthales, 170, 171
chordates, 105
Clonal interference, 4, 5
Cnidaria, 117, 118, 127, 135, 148, 149
  Anthozoa, 117, 118, 121, 126, 127
  Cubozoa, 117
  Hydra magnipapillata, 119, 131
  Hydractinia echinatta, 127
  Hydra viridissima, 131
  Hydrozoa, 117, 118, 121, 126, 127
  Nematostella vectensis, 121, 122, 130, 131
  Scyphozoa, 117
cnidocil apparatus, 137, 138, 144, 145
CnNk-2. Nkx-2.5, 121

Coelenterate, 120
Coelenteron, 120
Communication service, in integrative systems, 89, 90
cooption, 100
Covarion models, 30, 31, 35, 40
Cryptomeria, 168
Cubozoa, 135
Cycle sequencing, 185
cytochrome c oxidase subunit 1 (COI), 184

Darwin, 105
Defecation reflex, 123
defense, 137, 138
Dengue, 204
desmoneme, 138, 139, 142
developmental drives, 104
developmental genetic programme, 99
Diabetes, 202
Didinium, 139
Differentiation process, in Foraminifera, 79
Diffuse nerve net, 117, 124–126
Diffusion, 117, 118, 120–123, 131
dinoflagellate, 139, 145
diversity, 181
dN/dS ratio, 205
Drosophila, 131
Duplication, 156, 161, 162

Enteric nervous system, 124–126
Epifagus, 166, 168, 174
Epifagus virginiana, 166, 174
Esophageal reflex, 123
Eudicots, 170, 171
EvoDevo, 96
Evolution, 61, 63, 65–67, 69
  endomitosis meiosis, 64, 65
  meiosis, 63–65
  polyploidy, 63, 65, 66, 69
  protists, 63, 64, 67
  protozoans, 63, 65
Evolutionary biology concepts, 209–214
Evolutionary medicine, 198
exclusive organelle, 138
extinction risk, 183
extinction vortex, 193
Extracellular matrix, 128, 129

Fisher's fundamental theorem, 4
Fitness effects, 3
  additive, 5, 6
  multiplicative, 6
Fixation probability, 4, 5

Foraminifera, 74
  aromorphoses in, 84
  classification scheme, on shell features, 75, 76
  differentiation process in, 79
  integration process in, 81
  phyletic lines of, 74, 75
  polymerization process in, 77
Four-Taxon Simulation, 33, 35, 38
fragmentation, 181
Functional annotation, 209–214
Functional constraints, 46
Functional innovation, 55
Functional shift, 210
  neofunctionalization, 210
  subfunctionalized, 210
fundamental biogenetic law, 95

Gastrovascular cavity, 120, 123, 130
Gene age, 45, 47, 49–51, 53, 55, 57
Gene birth, 56
Gene duplication, 56, 57
gene loss, 148, 149
genetic diversity, 181
genetically depauperate, 193
Genome-wide association studies, 201
Girsanov transform, 3, 7, 8, 14, 17
Global Warming, 203
Group II intron maturase, 165–169, 173–176
  reverse-transcriptase, 168, 169, 175
  RT0 and RT3, 175
Gymnosperms, 168

Haeckel, 105
haplotypes, 186, 187
hemoglobinopathies, 199
Heterotachy, 29–31, 33, 35, 37, 38
HLA, 201
holotrichous isorhiza, 138
homology, 98
Human activities material objects, evolutionary development of, 81, 89, 90
Human society, development of, 86
Huperzia, 168
Hydra, 117, 119–123, 126–131, 135–147, 149
  basaldisc, 117, 128–130
  diffuse nerve net, 117, 124–126
  digestive tract, 117, 120, 122
  hypostome, 118, 123, 126
  nerve-freehydra, 124
  regeneration, 118
Hydra matrix metalloprotease, 128, 129
Hydrostat, 121, 122

Index 219

Hydrozoa, 135
hygiene hypothesis, 202

Illness, 198
Immune response functions, 52, 55, 57
inbreeding, 194
induction, 98
INFA, 173
Inflammation, 205
inflammatory bowel diseases, 201
inherency, 96
Integrative system in Foraminifera, 81–84
interleukin-10, 202
intermediate filament, 145
invertebrates, 181

Jellyfish, 120, 130

Language and human society
       development, 86
lateral gene transfer, 148, 149
law of embryonic divergence, 104
Life-cycle, 66–68
lifespan, 205
Lifestyles, 198
Likelihood, 29–38, 40
Lineage-specific adaptations, 50
Linkage disequilibrium, 4, 6
Litonotus, 139

Macroevolution, 62, 63, 65, 66, 68, 69
Magnoliaceae, 170, 171
Malaria, 204
Marchantia, 168
Martingale decomposition, 10, 12
Martingale problem, 12, 13
matK, 165, 167, 171–173
   Adiantum capillus-veneris, 166, 174
   domain X, 166, 168, 169, 173–175
   RT domain, 168, 169, 175
   zinc-finger, 169, 175
MATR, 173
Medicine, 198
Mesostigma, 169
Metagenomic approach, 213
Metamorphosis, 127
Microevolution, 61, 62, 65, 66, 68, 69
minicollagen, 146, 147
mitochondrial respiration, 204
Mixture Model, 29, 31–34, 37, 38, 40
Modularity, 98
Moment equations, 14, 17
Monocots, 170, 171, 175
Moran models, 6

Mortality, 204
Muller's ratchet, 6
multiple sclerosis, 201
Mustard, 173, 175

Negative selection, 46, 56, 57
nematocyst discharge, 138, 142, 143
nemertean, 148
neoDarwinian paradigm, 101
nested clade analysis (NCA), 186
Neutral process, 14, 17, 18, 20, 22
New head hypothesis, 105
Non-coding sequences, 57
Nonsynonymous, 166, 167, 172, 173
Novikov condition, 17, 18
Nucleotide substitution rates, 45
   non-synonymous, 46, 49, 56
   synonymous, 45, 49, 50, 56
Nymphaeales, 170

Oat, 175
Obesity, 202
operculum, 138, 141, 142, 144–147
Orchidaceae, 167
organizational homologies, 102
Orphan genes, 46, 50–52, 55–58
Orthologous, 46–49

p53, 66, 68
Paralog, 155–157, 160, 161
Paramecium, 138, 139, 142
Peduncle, 119–121, 130
Peristalsis, 122, 124, 126
Peristaltic reflex, 123
phenotype–environment curve, 199
philopatric, 183
Phyllocladus, 168
Phylogenetic reconstruction, 29, 30
phylogeographic inferences, 187
Physcomitrella, 168
Pinus, 168
Planula, 127
Plastic responses, 198
poly(γ-glutamate), 138
Polymerase chain reaction (PCR), 184, 185
Polymerization process, in Foraminifera, 77
Polyploidy, 63, 162
   endomitosis meiosis, 65
   protozoans, 63, 65, 67
polytomy, 188, 192
Positional information, 117, 118
Positive selection, 46, 50, 55, 56, 211
   selective pressure, 211
Potato, 175

pre-Mendelian world, 102
prey capture, 135, 137, 138, 142, 144
Primate-specific genes, 45, 47, 51–54, 57
Protein length, 45, 48, 51, 57
Proterozoic Era, 121
Protista, 73, 77, 79, 81
protochordates, 105
Protozoan, 61, 63, 65–68
Protozoan life cycle, 61, 66, 67
pseudocolony, 139, 140
Pseudogenes, 155, 158, 159, 161
Psilotum, 168

Rate of adaptation, 3
   asymptotic limit, 5
   upper limit, 4
rbcL, 166, 167, 170, 172
reaction norm, 198
reciprocal monophyly, 193
Remanella, 139
Reticular formation, 124
RFamide, 120, 121
Rice, 167, 168, 175
risk factors, 205
RNA-lysine, 165
rpl2, 174, 176
rps12 cis, 176
rps16, 175
rRNA, 155, 156, 158, 160, 162

Scyphozoa, 135
Sea anemones, 118, 120, 121, 126, 130
Selection coefficient, 3–5, 8, 11, 23, 25, 26
Selective sweep, 4, 5
senescence, 205
sequence divergences, 188
Sequence similarity searches, 46, 48, 53, 54
Sequiadendron, 168
sickle cell disease, 199
Sinapsis alba, 173
Social groups, in human society development, 86, 87
Social integrative system, material objects of, 89
spine, 141, 142, 144, 147

State and law systems development, 87
Stem cell, 61–63, 68
stenotele, 138–141
Strong selection, 3, 7–9, 11, 14, 23, 26
stylet, 138–144, 148
subfunctionalization, 162
Sugarcane, 175
susceptibility genes, 201
symbiosis, 149

taeniocyst-nematocyst, 139
Taxus, 168
terminal addition, 96
theory of heterogeneous generation, 96
Thrifty Gene Hypothesis, 202
tissue-specific, 160, 162
Tobacco, 167, 173
toxicyst, 138, 139
toxin, 138, 142, 144
Transition, 166, 168
translocation, 193
Transversion, 166, 168
Travelling Wave, 5–8, 14
trichocyst, 138, 139, 142
trnA, 176
trnG, 175, 176
trnI, 176
trnK, 165–167, 169, 173–176
trnL-trnF, 172
tuberculosis, 200
tubule, 138–142
tubulin, 145
Tumour, 61, 62, 65, 66, 68, 69
   meiosis, 61, 65, 66
   microevolution, 61, 62, 65, 66, 68, 69
   paleogenetic, 68

uncoupling, 204

Variance of fitness, 4, 5, 23, 25

Weak selection, 7, 17
Weak selection model, 3, 7, 8, 11, 17, 23, 26
Wnt-3a, 127, 128